HVAC **Control Systems**

HVAC Control Systems

Raymond K. Schneider
New York City Technical College

John Wiley & Sons   New York   Chichester   Brisbane   Toronto

To Karen, Jahnna, and Damon

Library of Congress Cataloging in Publication Data:

Schneider, Raymond K
  HVAC control systems.

  Includes index.
  1. Heating—Control. 2. Air conditioning—Control.
I Title.
TH7466.5.S35   697   80-23588
ISBN 0-471-05180-2

Printed in the United States of America

10 9 8 7 6 5 4 3 2 1

Preface

The person who understands controls is the most important individual in the HVAC field. My own on-the-job experience and discussions with many engineers, mechanical contractors, and service managers have convinced me of the truth of this statement.

This book was written to provide the kind of information needed by the person who wants to become proficient in the field of HVAC controls. No book can provide such proficiency. It is learned on the job in the day-to-day activity of designing, installing, operating, and repairing such systems. A book can, however, provide the foundation upon which the practitioner can build. In this book will be found the vocabulary, the diagrams, the principles, and the philosophy that should give one the confidence to enter the field of HVAC controls. The reader will obtain an overview of the subject as well as a specific coverage of the hardware and systems in use today to provide for human comfort.

This book is intended as a vehicle for learning. Whether the reader is a student involved in a two-year community college technology course or a four-year engineering program or is a journeyman in the field seeking a key to unlock the mystery of controls, there is much information here that will be useful. If the course of instruction is in the form of a lecture, the text can serve as a resource. If the course is discussion-oriented, the Discussion Topics at the end of each chapter serve as a means of getting the students involved. A design course in HVAC controls will profit through the use of the wiring diagrams and design problems contained in the appendixes. The many photographs and sequences of operation included will further the aims of a controls laboratory program and be of inestimable use to the person working in the field.

The presentation of subject material is perhaps somewhat unusual. Early in the text the focus is general. The student is presented with the objectives of HVAC systems and the philosophy of how and why complete systems are designed. As the book progresses, the focus turns to specific hardware and how the objectives of the systems are achieved. Only one complete chapter is devoted to troubleshooting, but numerous suggestions are given throughout the book on potential problems and how to prevent or solve them.

Although much of the information contained in this book is the result of my own activity in the field, I would like to acknowledge the significant help offered by the firms in our industry. Many of the photographs and diagrams that bring the words to life were willingly contributed by these companies. A word of thanks also goes to my colleagues across the country whose comments during the review stage of writing helped in sharpening the focus of the book.

Raymond K. Schneider

Contents

Chapter One Introduction

Control. It is the intent of this book to give meaning to this word as it relates to the heating, ventilating, and air conditioning (HVAC) of buildings. The term *HVAC* has been applied to six functions that are controlled in order to meet environmental requirements—heating, cooling, humidification, dehumidification, air movement, and air cleanliness. Although there has been considerable effort to gather these terms together under the term air conditioning, the fact is that HVAC is the common expression in use today, with air conditioning being generally understood to apply only to the cooling function.

The reason for the existence of any HVAC system is to provide control of a particular environment. The environment may be for humans at work, or resting in their homes, or shopping in a store. The environment may be that in which an industrial process takes place, enabling machinery to work smoothly or permitting plants to grow. The environment may be that of a hospital, or a school, or a bank—the list is almost endless. There is no precise way to group the different environments that must be controlled, but the expression *applications* is in wide use in the following areas.

1. *Commercial applications* refer to an HVAC system that controls the environment of stores, offices, banks, restaurants, and generally those places where people come and go,
2. *Residential applications* refer to the environmental control of homes, apartments, condominiums, and similar places where people live, sleep, and cook.
3. *Industrial applications* refer to controlling the environment for industrial processes, storage, computer operations, assembly areas, and the like.

When we talk about a particular HVAC application, we are really considering how principles of control of the six basic functions can be *applied* to the particular *requirements* of an environment. Although three broad applications were previously mentioned, there are in fact many more, and any effort to categorize them precisely is a waste of effort. The American Society of Heating, Refrigerating, and Air Conditioning Engineers (ASHRAE), in their *ASHRAE Handbook on Applications,* points out the uniqueness of various environmental requirements by dealing with dozens of specific applications of HVAC control principles. Although much of the information contained in this book will deal with specific applications, you will find that much more of that information is general enough to apply across a wide range of applications.

1.1 FUNCTION OF A BUILDING

Buildings are erected with a purpose in mind. In many cases the exact purpose of a building may not be known; such a building is usually built on "speculation." The owner is counting on the building's being applicable to the needs of some unknown tenant at some point in the future. A building of this kind may house offices or storage space or perhaps a manufacturing facility or a retail establishment. The owner is "speculating" that although a specific need has not been identified at the time of design, a tenant will be located by the time the building is ready for occupancy.

The alternative to a speculative building is one built with a definite purpose in mind. A factory, an executive office building, a school, and a restaurant are examples of buildings usually designed for a definite purpose; from design through building that purpose is known.

The function of the building is usually not decorative—it should blend in with its surroundings and enable the activity for which it was planned to be conducted in an efficient manner. We define *efficiency* as the ratio of output to input, but a mathematical calculation of the efficiency of a building, practically speaking, is impossible to determine. We can, however, look at a building in an overall sense and get a feeling for the way the HVAC system will affect the activity within.

The control that we wish to have over the HVAC system is directly related to the function of the building. Whether you are involved in the design phase of the system and want to find guides as to which design course to take, or in the maintenance and repair phase and want to know the reasons for designing a system a particular way, it is useful to think about the function of the building.

The HVAC of a speculative building must be flexible because we do not know its final purpose. It must be economical because speculative investments are risky and the less invested the better. It must be sophisticated to attract quality tenants with high requirements. It must be simple, hence inexpensive, to attract tenants with limited budgets. It must incorporate extensive energy saving equipment to reduce the operating cost of the speculator (the owner). It must minimize the use of sophisticated control systems so that, if the speculator decides to sell the building, he can keep his price down in a buyer's market.

If you have spotted some contradictions in the preceding discussion, you can begin to understand some of the problems offered by speculative building HVAC systems.

Buildings with a definite purpose, on the other hand, are quite straight forward. The HVAC system can be inflexible, since we know its intended use exactly, yet it should also be flexible to accommodate change and the expansion of its intended activity. It should be simple enough to facilitate maintenance but sufficiently sophisticated to meet exacting requirements. It should minimize energy consumption, yet be inexpensive enough to meet the building construction budget.

At this point it should be quite clear that HVAC control is not an exact science. There is no one solution for a particular application.

As we proceed through the rest of the book, a number of different techniques will be described. Which techniques should be selected will depend on such factors as the available funding, equipment shipping dates, power supply availability, building codes, or other factors not directly related to the science of HVAC controls as such.

1.2 CONTROLS AS AN HVAC SPECIALTY

In erecting any kind of building, a number of skills are required. In recent years the HVAC requirements of large buildings have become quite complex and require people uniquely skilled in this area. In small residential or commercial applications the air conditioning contractor or electrician is usually responsible for controls along with other responsibilities. In large buildings the specification for the HVAC system usually has a separate section for controls and, in fact, may often specifically require that people specializing only in control systems install and check out the controls.

The *HVAC engineer* describes how a system should be controlled. The *controls designer* selects the equipment that will accomplish the intent of the HVAC engineer and then diagrams it for installation. The *controls draftsman* formalizes the plan of the designer and develops the control drawings used to communicate with the personnel in the field. The *controls installer* will install the controls in accordance with the manufacturer's instructions and the controls drawings. The *controls serviceman* will start up the system and insure that the final control system does what the engineer originally specified. The controls serviceman also maintains and troubleshoots systems once they are in operation to insure that they work properly over the life of the equipment.

The person who succeeds in the controls field is one well versed in all types of HVAC systems. The controls person must be able to identify the components of the HVAC system and know how they are intended to interact. Such a person must know how to read diagrams that describe how systems work, the sequence of operations, and the manufacturer's operating instructions. Then, after digesting all the information, the controls person must be able to come up with a plan of action, which may be in the area of design, drafting, installation, service, or maintenance.

1.3 CONTROL VARIABLES

Our HVAC control system is intended to maintain a set of environmental conditions in a space, automatically. This is accomplished when the *control variables* of the space are kept at the values we specify.

The control *variables* are measurable quantities such as *temperature, humidity, velocity,* and *pressure*. Although we are looking for human comfort in a residential HVAC system, there are no instruments that will give a measure of comfort. In a computer room we want the equipment to run trouble-free, yet we have no instrument that will give us a direct indication that the environment in the computer room will permit such operation. What is necessary is a specification of control variable values that will produce the desired effect.

The HVAC applications mentioned earlier differ primarily in the way we specify control variables. A residential application may require a minimum temperature of 72°F, using heating only. If it requires a minimum of 72°F and a maximum of 78°F, the HVAC system must be capable of both heating *and* cooling. An additional specification of 50% relative humidity would require an additional humidification and/or dehumidification capability. It should be apparent at this point that the greater the number of variables we want to control, the more sophisticated the equipment and control system and, therefore, the more expensive it will be.

A real challenge that must be faced by the control engineer is to obtain an understanding of the function of the building so that he or she can specify the appropriate equipment to control the variables that must be controlled to satisfy the requirements of the building.

Sources of Design Information

The *ASHRAE Handbook of Applications* is probably the most widely recognized and accepted source book. In its pages one can find recommended design temperatures, humidity levels, pressures, types of equipment and control systems, cost considerations, and a wealth of additional information that will help in developing a total specification.

In some cases, a building owner with experience in a particular field will specifically detail the value of the control variables and will give the design engineer little or no leeway in this area.

Another source of design information is the trade association of a particular industry, which often has technical information available or can refer one to a recognized expert in the field.

Once a determination has been made about which variables will be controlled, the next task is to determine how accurately they must be controlled. In a residential application, a temperature deviation of more than 1½ to 2°F from the desired temperature may cause discomfort. In a store, however, where people move around and may be wearing outer garments, the temperature may be allowed to deviate as much as 3 or 4°F. In an industrial process, temperature changes of more than two- or three-tenths of a degree may be unacceptable. It should be apparent that the more precisely we attempt to control the variables of a system, the more complex and expensive the HVAC system will be.

Many HVAC systems are in operation today that are either incapable of meeting application specifications or are more complex than necessary to do the job. A very elaborate control system connected to very simple equipment is just as much a waste as a very simple control system connected to sophisticated equipment.

In the following pages we shall encounter equipment that is both simple and complex along with controls for each kind. A designer will want to be able to identify the capability of a control system in order to match it to the equipment and the application. A service person will want to identify the capability of the equipment and controls so that he or she will know just how much to expect from the system being serviced.

1.4 THE PURPOSE OF CONTROLS

Why must we have control? We must control the HVAC system in order for it to do the job we desire, namely to provide an environment that enables a building to function efficiently. The HVAC control system monitors the control variables and regulates or adjusts the performance of the equipment to maintain the desired environmental conditions. This is the primary purpose of the controls, which are generally referred to as *primary controls*.

A secondary purpose, and some might argue about its being secondary, is safety. Much of the HVAC equipment we encounter is potentially hazardous. It is the purpose of *secondary controls* to regulate the operation of the system so that it is efficient, and more important, safe.

In our real world the control of HVAC equipment is not left totally in the hands of the control designer. Building codes have been developed to offer guidance in the way HVAC systems are to be designed and installed. Although a seemingly new and revolutionary technique of control or energy conservation may be lurking in the mind of the control designer, it must wait to be proved and accepted by the world. Such proof and acceptance is by means of "the code."

There is no uniform building code across the country. Typically, codes are developed on a local level; they govern the manner in which buildings are erected and maintained in a city or county or other political subdivision. In most cases, however, the codes are not totally unrelated to other codes around the country. They are related by requiring compliance with standards developed by generally recognized organizations. Most codes specify that electrical equipment must comply with UL (Underwriters Laboratories) Standards, that gas-fired equipment must be listed with AGA (American Gas Association), that boilers must be listed with IBR (Institute of Boiler and Radiator Manufacturers), or that the fire detection system must be in accordance with NFPA (National Fire Protection Association) Standards. Whereas local codes may have a few provisions that are unique to a particular locale, in general, they have a greater degree of uniformity than might be expected.

The control designer must not only be aware of how HVAC systems can be controlled but must also be aware of how they *must* and *must not* be controlled to meet code requirements. The purpose of control is to do the job, do it safely, and meet the code requirements.

1.5 METHODS OF CONTROL

We can consider a control system to be composed of two parts: (1) a *controller* that has the ability to sense a control variable and determine whether or not it meets our design specification, and then send out a signal or directive to the HVAC equipment to take corrective action if necessary, and (2) a *controlled device* that receives the directions from the controller and implements those directions.

A wall-mounted thermostat found in a home is an example of a controller. It analyzes the room temperature; if the temperature is too low, it sends a direction to a gas valve in a warm air furnace for it to open. The gas valve opens, gas flows and is ignited, and heat pours into the room a minute or two later.

As the heat in the room increases, it will cause the thermostat to sense an increase in temperature and eventually send a direction to the gas valve to close, stopping the flow of heat to the room. This type of control system, in which the controller senses the result of its action is called a *closed loop* control system.

An *open loop* system is one that does not incorporate *feedback,* that is, one in which the controller does not sense the result of its action. As an example, a thermostat mounted on an outside wall of a building senses the outdoor temperature. When the outdoor

FIGURE 1–1   This wall-mounted thermostat used in home heating applications has a maximum heating set point of 74° F. This so-called "energy conservation" thermostat replaces similar appearing thermostats with maximum settings as high as 90°F. (Courtesy of Honeywell Inc.)

temperature drops to, let us say 50°F, the thermostat is supposed to send directions to the boiler to start up and begin providing hot water to the building's heating system. Now this outdoor thermostat does not have feedback, that is, it never feels the hot water for which it is calling, it only senses the outdoor temperature. Because it does not sense the result of its action, it is part of an open loop control system.

It turns out that complete control systems of increasing complexity are really combinations of open and closed loop systems arranged in a variety of fascinating ways.

As just stated, the controller sends out directions. Just what form do these directions take? While a tenant banging on a radiator demanding more heat from the landlord because the tenant is cold is an example of a control system (tenant = controller, landlord = controlled device) the typical direction of a controller is nonverbal. As we shall explore in depth later, typical directions are in the form of electrical, pneumatic, or electronic signals. Before we get too deeply involved in each of these specific forms of signal, we ought to look more closely at how the signal will attempt to achieve its purpose, that is, cause the HVAC equipment to control the environment. The directions of the controller for corrective action can be formed in several ways, which we call *control action*.

1.6 CONTROL ACTION   The wall-mounted home thermostat mentioned previously is a good example of a *two-position* control, the first of five control actions that we shall discuss. The direction sent out by this controller is either "send heat" or "send no heat." The gas valve is either fully open or fully closed and, as a result, the furnace is either operating at full heating capacity or at zero heating capacity.

This "two-position" control action can be diagrammed as shown in Figure 1–2. The vertical axis of the graph is the controlled variable, in this case *temperature;* the horizontal axis is *time.* A close look at this diagram should make the understanding of other control actions a bit easier.

At some point, as the temperature is falling, the thermostat calls for heat—that is labeled "on". The temperature continues to fall a bit more because it takes a few minutes from the time the thermostat directs the gas valve to open until the heat is actually felt in the room. This "lag" depends on the size of the building, the distance between the furnace room and the thermostat, how cold it is outside, and other factors. In any case, the temperature finally begins to climb and will continue to climb until the thermostat directs the gas valve to close. That is labeled "off" on the diagram.

Note that the temperature continues to climb a bit before it levels off and starts to fall. This is called "overshoot" and is a result of a buildup of heat in the ductwork and furnace that must be removed before the temperature starts to drop. The difference between the "on" and "off" points of the thermostat is called the *control differential.* The difference between the high temperature and low temperature in the room is called the *operating differential.* Ideally, we would like them to be the same, but in reality they are somewhat different. Generally speaking, the smaller the operating differential of the system, the better the system. As we look at other control action we shall see that much of the sophistication added to this simple "on-off" two-position control is simply for the purpose of reducing the operating differential of the system.

One way that the operating differential can be reduced is by employing the second of the control actions we shall consider, namely, *timed two-position control.* In this control action a small

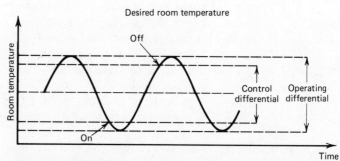

FIGURE 1–2   The control differential is built into the thermostat by the manufacturer whereas the operating differential is related to the heating system design, building construction, and other factors. A typical residential thermostat may have a control differential of 1½ to 2° F, but thermostats used in commercial applications may have differentials as high as 4 to 6° F.

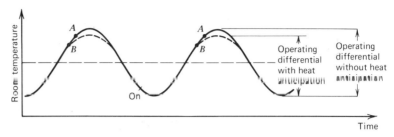

FIGURE 1–3   The solid line shows a normal two-position control curve with point *A* being the point at which the gas valve shuts off. In timed two-position control the valve would shut off at point *B*. The room temperature does not rise to as high a level, reducing the operating differential. Note that this action does not affect the lag, only the overshoot.

heater is built into the thermostat. When the heating thermostat calls for heat, the small heater is energized along with the gas valve. The heat generated within the thermostat tends to cause the thermostat to close the gas valve sooner than it would otherwise. The overshoot of the system still causes the temperature to rise a bit after the gas valve is closed, but it does not rise as much as it would normally. Figure 1–3 illustrates this control action, also called "heat anticipation," which reduces the operating differential. This same anticipation action can be obtained in cooling thermostats by energizing the heater during the "off" period, causing a false or premature cooling demand.

Both *two-position* and *timed two-position* control action produce full-on or full-off positioning of the controlled device, the gas valve in the example given. In HVAC systems, where the controlled variable changes rapidly upon signal from the controller, a third control action can be considered, that is, *floating action*.

The controller in a floating control application has three positions rather than the two positions of the controls mentioned previously. The third position is a neutral or "dead band." The heating controller sends a signal asking for heat, for no heat, and also a third signal that freezes the controlled device at some intermediate position. The controlled device has to be designed to respond to such a signal.

In Figure 1–4 floating control is diagrammed. As the temperature falls, the thermostat, designed for floating control, will send out a signal for heat (point *A*). The gas valve, designed to open on a call for heat, to close on a call for no heat, and to stop at an intermediate position when it receives a neutral signal, *gradually* starts to open. As it slowly moves toward the open position, the heat output of the furnace gradually increases, and, if we were looking at the gas flame, we would actually see it gradually increasing in height.

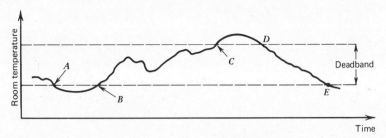

FIGURE 1–4 The floating controller sends a signal to the gas valve causing it to move toward an open position between *A* and *B*. The valve stops moving between *B* and *C* and begins to close between *C* and *D* stopping again as the temperature dips down into the "deadband" between *D* and *E*.

Since the furnace output is increasing, we might see the temperature of the air striking the thermostat increase before the valve is fully open. If the temperature increases sufficiently, the controller senses that increase and sends a signal to the valve that stops it in a partly opened position (point *B*). At that point the temperature is free to "float" uncontrolled within the limits or differential of the controller. A constant amount of heat is being added to the air, since the valve is partially open. If this amount is too much, the temperature moves upward until the thermostat senses that the gas valve should be moved toward a closed position (point *C*). Perhaps the valve fully closes or perhaps assumes another position allowing less gas to flow and therefore reducing the output of the furnace. If the heat output is inadequate, the temperature falls (point *E*) and the thermostat directs that the gas valve open still farther.

In two-position action the controlled device (gas valve) moves to a full-open or full-closed position. In floating action the controlled device (gas valve) assumes an intermediate position only when the controlled variable (temperature) is within the differential of the control. In *proportional action* the controlled device assumes a

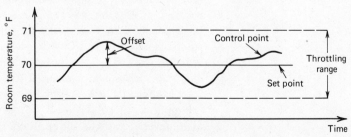

FIGURE 1–5 Proportional control and proportional control with reset action. The major difference is that the offset tends to be less in the system incorporating reset action because such systems attempt to return the control point to the set point.

position directly related to the difference between the temperature desired and the actual temperature sensed by the controller.

Figure 1–5 is somewhat similar to the diagrams presented previously, but it introduces some new terms directly related to proportional control action. The *set point* is the temperature we would like to maintain in a conditioned space, let us assume 70°F. The *control point* is the actual temperature in the room. The *offset* is the difference between what we want, the set point, and what we have, the control point. It follows then that the greater the offset, the greater the requirement for corrective action by the HVAC system.

Proportional action is action in which the controlled device assumes a position proportional to the offset. In our gas-fired furnace example we need a thermostat specifically designed for proportional control. We assume a throttling range of 2°F. That is, the valve will go from full-open to full-closed, maximum heat output to zero heat output, as the temperature changes 2°F. So if the set point is 70°F, the valve will be fully open at 69°F and fully closed at 71°F. At 70°F the valve will be half-open; at 70.5°F the valve will be one-quarter open; and at 69.5°F the valve will be three-quarters open. The valve position is *proportional* to the deviation of the room temperature from the desired temperature.

In proportional action systems it is more or less an accident when the control point and the set point coincide. The positioning of the gas valve tends to maintain the offset, whatever it might be, as long as the heating requirement is constant. If the room gets a bit cooler, the valve will open a bit more, but it will not provide excess heat to increase the temperature and bring it back to the set point. Rather it will just provide sufficient increased heating output to maintain the new control point. When the temperature drops below the throttling range (or goes above it), the system is said to be out of control.

Proportional action with automatic reset is intended to provide positive action in getting the room temperature back to the set point. This fifth and final action is the most capable of holding a set temperature with a minimum deviation of all the actions described. In effect, a controlled device *overreacts* to a change in the control variable sensed by the controller. If the thermostat of the previous example sensed a decrease below the set point of 0.5°F down to 69.5°F, it would send a signal to the gas valve causing it to move to a ¾ open position. The heat output of the furnace would match the heating requirement and the temperature would stabilize at 69.5°F.

If the thermostat were equipped with an automatic reset, however, it might move the valve to a ⅞ open position. The furnace would generate more heat than was required to hold 69.5°F. The result would be that the temperature would move upward, toward the set point. The intended effect of this action is to decrease the operating differential of the system by continually attempting to maintain a set temperature.

The control actions described are found in various applications in the HVAC field. The remainder of the book is devoted to describing how these actions are applied to primary source applications such as compressors, gas burners, fans, and oil-fired boilers, and to distribution systems such as central fan systems and hot water systems with their accompanying terminal equipment.

DISCUSSION TOPICS

1. Define the term *application*. Give several examples of applications and discuss their similarities and differences.
2. How might the function of a building affect the design of the HVAC system?
3. What is a control variable? Give several examples.
4. How might local building codes be similar and how might they be different from one locale to the next?
5. Describe an open loop control system and a closed loop control system. Which is superior and why?
6. Why is timed two-position control considered an improvement on two-position control?
7. How might floating control prove to be an energy-saving device?
8. Why might proportional control with automatic reset be considered superior to proportional control?

Chapter Two   Producing Useful Heat

Having looked at an overview of the control function, we can now consider some of the details. Our first efforts in controlling the environment were in the area of heating. Until rather recently, this effort was restricted to building a wood fire and regulating the output of the fire by adding logs more or less rapidly as comfort dictated. Although woodburning equipment has recently enjoyed a remarkable resurgence in rural and woodland areas, it has been almost totally replaced by coal-, oil-, and gas-fired systems.

Of these three popular sources of heat, coal is most widely used in the area of power generation while oil and gas find application primarily in comfort heating. The major thrust of this chapter will be in describing the controls used in producing useful heat from oil and gas. Incidentally, electricity as a direct source of heat energy is not insignificant. Its use was growing rapidly until the fuel crunch of the early 1970s resulted in rapidly escalating electricity costs. The popularity of electric heat will increase again once the problem of finding a source of electricity—nuclear, solar, or wind—is solved. Controlling electric heat will be mentioned toward the end of this chapter.

2.1 ENERGY CONVERSION

The energy conversion process of interest in comfort heating is one in which chemical energy, provided by gas and oil, is converted to heat energy. This process, called *combustion,* must be controlled, primarily to provide the amount of heat we want, and secondarily, to provide heat safely and efficiently.

The combustion process occurs in the *combustion chamber;* the hot products of combustion pass through a *heat exchanger,* where heat is extracted from the hot combustion products or flue gases; and, finally, the now relatively cool gases flow up a stack or chimney and are released to the atmosphere. The process of extracting heat from the flue gases and distributing it to the

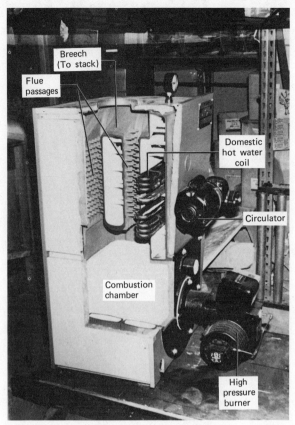

Breech
(To stack)

Flue
passages

Domestic
hot water
coil

Circulator

Combustion
chamber

High
pressure
burner

FIGURE 2–1   This cutaway boiler shows the combustion chamber and flue passages through which the hot products of combustion flow on the way to the stack. The high pressure gun-type burner mounted on the front of the boiler mixes oil and air and introduces the mixture into the combustion chamber where ignition takes place. Note the coil mounted inside the boiler that produces domestic hot water year round. The circulator mounted above the burner is also included as part of this package.

conditioned space will be covered in the next chapter. For the present we are concerned only with controlling the combustion process itself.

When our ancestors sensed that they needed heat, they started a fire with a match, flint, or some other means. When our automatic heating control system senses that we need heat, it will also start a fire. This action is called *ignition*.

As our ancestors felt the need for more or less heat, they added logs to the fire at a rate that would keep them comfortable, or at least reasonably comfortable. Today's heating systems also have the capability of adding fuel to the fire in varying amounts from zero to a maximum based on the size of the equipment. Depending on the application we can expect to see any of the control techniques described in the previous chapter used in controlling the rate of combustion.

When our ancestors saw that the fire was getting out of hand, they put it out or took other corrective action to cut it back to a safe level. Today's systems are equipped with safety devices or *limit controls* that observe and take corrective action automatically.

The reason for describing our present rather complicated heating systems in terms of the old woodburning fireplace is to impress upon the student the fact that today's systems did not spring out of an engineer's head in the blink of an eye. Rather they evolved slowly over a long period of time. The computer-operated, total environmental control systems of today are based on systems installed 10 years ago, which in turn were modifications of systems installed 10 years prior to that, which were based on still older systems that eventually lead us back to that ancestor in the cave. Controlling the combustion process today means simply providing ignition, fuel control, and safety limits, just as it always has been, except that today it is done automatically, reliably, efficiently, cleanly, and usually on a much larger scale.

2.2 IGNITION

Once a determination has been made that heat is required, flame and fuel must be brought together to provide ignition. Somehow the fire must be started. Two techniques are in widespread use: (1) the use of an electric spark, usually generated by a transformer with a 6000-to-10,000-volt output, and (2) the use of a small fuel burner, called a pilot, already lighted either manually or by electric spark, to ignite the main burner. The term *burner* in this discussion refers to the device that directly feeds air and fuel, whether gas or oil, into the combustion chamber.

In small residential heating systems the ignition is rather simple. The spark ignition technique is commonly used in oil-fired systems. On a call for heat by the main thermostat, oil and combustion air begin to flow and the transformer establishes a high voltage spark in the path of the oil/air mixture. Systems having the spark arcing continuously as long as the fuel flows are called *continuous ignition* systems; those in which the spark stops after ignition has occurred are called *intermittent ignition* systems.

Most smaller gas-fired heating systems in operation today use a standing pilot, that is, a small gas flame that has been manually lit and burns continuously. On a call for heat by the main thermostat, gas flows through the burner, comes in contact with the pilot flame, and ignites.

In recent years, with the increasing awareness of the short supply of gas, there has been a trend away from the standing pilot, which has been estimated to consume as much as 8% of all the gas used in cooking, hot water heating, and space heating. It is being replaced by spark ignition systems similar to those used in oil-fired systems.

In larger equipment a combination of both techniques is often seen. One version has a spark igniter lighting a gas pilot, which then ignites the main burners, which could be oil- or gas-fired. Another version uses a spark igniter to light a small oil burner using light domestic fuel oil (No. 2 oil), which then ignites the main burner which uses a heavy fuel oil (No. 6 oil).

2.3 FUEL CONTROL

The most common method of controlling heat output in small residential applications is by "two-position" control. On a signal from the space thermostat calling for heat, fuel oil or gas flows, ignition occurs, and heat output is constant at a maximum value. When the space temperature is satisfied, the fuel flow is reduced to zero as is the heat output.

In small oil-fired systems an electrically driven pump controls the fuel flow; the positions are either "on" or "off." In some

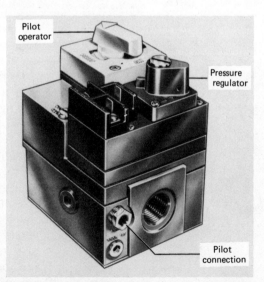

FIGURE 2–2   The valve shown at the left is a gas valve energized by a 24-volt signal from the thermostat. At the right is a combination valve that incorporates a pilot safety and a gas pressure regulator. Combination valves are commonly found on smaller residential systems; the standard gas valve finds application in both small and large systems. (Courtesy of Honeywell Inc.)

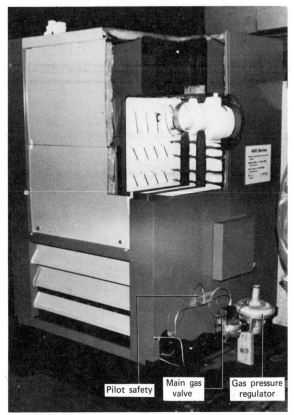

Pilot safety | Main gas valve | Gas pressure regulator

FIGURE 2–3   A gas-fired boiler rated in hundreds of thousands of Btu per hour. Note, at the lower front of the boiler, the gas pressure regulator, gas valve, and pilot safety. In smaller boilers these are combined in a single valve called a combination gas valve.

systems an electrically operated valve, called a solenoid valve, is used to insure complete instantaneous shut-off.

A gas-fired system is quite similar in that it has a two-position valve, which responds to the main thermostat. Although solenoid valves are used, in which an electromagnetic force is created to open the valve, other designs of valves are also encountered. An example is a heat-operated valve in which an electric current flows through a tiny heater and causes a vapor charge to expand forcing the gas valve to open slowly. Such a valve is characterized by a time delay of as much as 30 seconds or more from the time the thermostat calls for heat and the gas valve opens. Another valve design uses a small solenoid mechanism to open a tiny port that enables gas to exert pressure against a diaphragm, causing the diaphragm to lift and permit gas to flow.

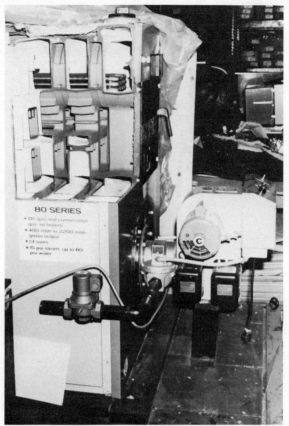

FIGURE 2-4   This combination boiler, available both in hot water and steam configurations, can generate heat using either gas or oil. In some localities the utility supplying gas charges a penalty if gas is used year round. Typically, on very cold days, or between certain fixed dates in the year, the owner will switch over to oil to avoid this penalty.

In an oil-fired system the pump establishes the oil pressure along with the nozzle selected for the burner. In a gas-fired system a pressure regulator is required to maintain the gas pressure entering the burner at the proper level. In many small residential applications this regulator is built into the gas valve and the name *combination valve* is applied to it. In larger systems the regulator is usually a separate piece of hardware.

In both the oil and gas systems described to this point, the flow of the fuel/air mixture was either at a maximum or at zero. As we look at larger applications we find that this is not always the case.

In larger commercial systems a common technique used to vary the heat output of gas-fired systems is called "staging." Two or more banks of gas burners are placed side by side and ignited one after the other as the heating demand increases. Each bank has its own gas valve energized by a thermostat in the conditioned space, which is designed to give two- or three-stage operation.

"Two-position with low fire start" is a phrase describing an intermediate-sized heating system. On a call for heat an amount of air and fuel equal to about 60% of full flow is introduced into the combustion chamber. After ignition of this mixture, the fuel and air volume is adjusted to full flow or "high fire" as it is called. The fuel mixture adjustment is usually made by means of a second valve piped in parallel with the first, which is energized along with an air damper used to control the volume of the combustion air. This damper is usually opened by an electric motor.

"High-low" control provides two steps or stages of heating output control. On a call for heat the system starts on "low fire" and after ignition automatically goes to "high fire" or maximum heat output. As the temperature approaches the desired temperature, or set point, the system throttles down to low fire, about 60%

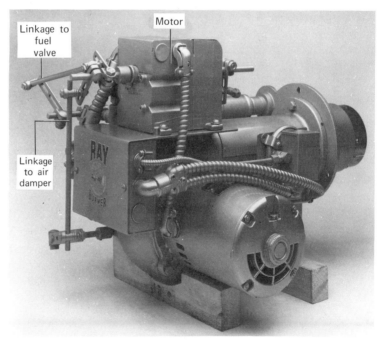

FIGURE 2–5   This burner is designed to fire either gas or fuel oil. Modulating control with a turndown ratio (ratio of maximum to minimum firing rate) of 3:1 can be achieved with appropriate controls. Note the damper motor and linkage that controls the combustion air supply at low fire and high fire and during modulation. This motor also varies the fuel firing rate. (The Ray JCE Combination Oil/Gas Burner, manufactured by Ray Burner Company, San Francisco, Calif.)

of maximum heat output. Once the desired temperature is reached, the system shuts down.

Modulating control provides for varying the heat output of the system by varying the amount of fuel and combustion air being introduced to the combustion chamber as the system demand varies. In large systems the "fine tuning" of the temperature in the conditioned space is not done in the combustion chamber but rather in the distribution system, as we shall see in the next chapter. Any modulation built into the combustion process, however, will usually increase the ability of the HVAC control system to maintain the required conditions and also add to the overall efficiency of the system.

2.4 SAFETY

Having an automatic capability of lighting a fire and a means of controlling the heat output of the fire is only part of the control system requirement. The final aspect to be considered is safety. Safety controls are intended to prevent damage to the heating equipment if the operating controls malfunction or are improperly set. They are also used to protect people and property. Safety controls can be considered to be of two general types: (1) *limits*, which do not operate unless an unsafe condition exists, and (2) *interlocks*, which are operating controls designed to ensure that components of an HVAC system are energized in proper sequence. Some specific control devices have built into them a combination of both limiting and interlocking action.

In large measure it is the safety controls that add to the complexity of HVAC control systems. In smaller heating systems the controls may be rather simple, sensing only that ignition has been accomplished when the thermostat calls for heat or that the equipment is not being overheated or subjected to undue pressures.

Larger equipment will have additional controls that insure that the oil temperature is proper, that the gas pressure is adequate, that the burner is properly positioned, that combustion airflow is as required, and that the water level in a boiler is adequate.

The variety of safety control configurations is so extensive that only a few typical examples can be presented here. A good controls person accumulates a library of manufacturer's literature describing the specifics of that manufacturer's equipment. One will find that there is enough similarity between the various manfacturers to enable one to adjust to differing equipment quickly once the basics are understood. Let us look at some examples of heating systems.

2.5 SMALL HEATING SYSTEMS—OIL-FIRED

Gas- and oil-fired heating systems in single or small multistory buildings are quite similar in principle, yet sufficiently different in specific configuration to warrant considering them separately.

First let us look at an oil-fired hot water boiler. The room thermostat, when it feels the need for heat, will send a signal to a relay, which in turn will energize a hot water circulator and the oil burner motor. A relay is an electrically operated switch and is commonly used in control systems where two or more different voltages are used. In the typical small application the thermostat has a 24-volt power supply while the burner motor uses 120 volts, hence the necessity of a relay. In effect the thermostat is telling the relay to turn on the oil burner motor and pump. When the burner comes on the combustion air fan and the fuel pump, both connected to the motor, force a combustible mixture of oil and air into the combustion chamber. This passes over an electric spark generated by the transformer, which was energized by the same relay as the burner motor. The spark causes the fuel/air mixture to ignite and the heating process begins.

At this point only operating controls are in action, that is, the room thermostat and the relay.

Suppose that the transformer had burned out and no spark was generated. There would be no ignition, yet the oil pump would continue to run, pumping oil into the combustion chamber. To prevent this problem, a control that provides "proof of ignition" is required. Two techniques are in common use today, the stack switch and the photocell. The stack switch is a thermostat that senses the temperature of the products of combustion in the stack (see Figure 2–6). If ignition occurs, this temperature will increase rapidly and the stack switch, sensing this, will permit the burner to continue to operate. Of course, if there is no ignition, the temperature will not rise. If it does not increase within 45 to 90 seconds, the switch will open and shut down the burner. If the flame is

FIGURE 2–6   The bimetal sensing probe of this "stack switch" is inserted in the stack to sense the flue gas temperature. Upon startup, if the stack temperature does not rise to an acceptable level within 70 seconds, the primary control, built into the body of the device, will lock out. A manual reset button is on the face of the control cover. (Courtesy of Honeywell Inc.)

Highlights—Oil Burner Control

The oil "primary control" has the purpose of providing switching for the burner motor and ignition transformer as well as monitoring the flame during burner operation. Figure 1 shows the wiring diagram for one type of primary control employing a cadmium sulfide cell.

SEQUENCE OF OPERATION

1. On a call for heat by the thermostat the circuit shown in Figure 1 is completed. Since the cad cell initially "sees" no flame, its resistance is very high (100,000 ohms and more) and it acts like an open switch.
2. As a result of the 1K relay coil being energized, contacts 1K1 and 1K2 close (see Figure 2), the burner motor begins to operate and the ignition spark across the electrodes in the burner is generated.
3. Upon ignition of the fuel/air mixture in the combustion chamber, the cad cell, sensing the flame, has a reduced resistance (100 to 300 ohms) and acts like a closed switch, energizing coil 2K (see Figure 3).
4. Coil 2K, when energized, causes the NO 2K1 contacts to close and the NC 2K1 contacts to open. This causes the current energizing the 1K coil to bypass the safety switch heater. At this point the burner will remain in continuous operation as long as the thermostat calls for heat.
5. If ignition had not taken place, the cad cell would have prevented coil 2K from being energized and the safety switch heater would remain energized. The heat generated by this heater would, after the designed time delay (say 30 seconds), cause the safety switch to open, deenergizing coil 1K, and the burner would shut down.
6. A similar sequence follows if flame failure occurred during burner operation. In both cases a manual reset switch must be depressed to get the system back into operation.

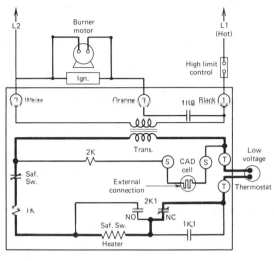

FIGURE 1

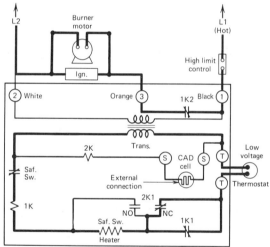

FIGURE 2

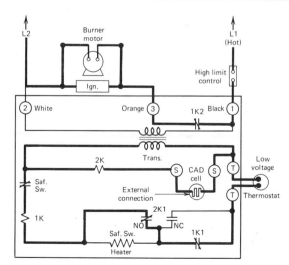

FIGURE 3

established initially, but at some time during the heating cycle the flame goes out, the same action will occur. This may be caused by an empty fuel tank. On some designs the stack switch may attempt to restart the burner once or twice, but if ignition is not proved it will then shut down until it is manually reset; that is, a button is pushed resetting the control.

The photocell is a device that is sensitive to visible light. It is mounted in such a way that it can peer into the combustion chamber and see if a flame is present. It works like the stack switch in that it will shut down the burner if ignition is not achieved within some short period of time. There are a couple of ways that photocells can be constructed. One of them uses cadmium sulfide as the material that is sensitive to light. The term *cad-cell* is widely used when referring to such a device.

Let us look at another problem. Suppose, with the burner operating at full capacity, that the amount of heat going into the water of the boiler is greater than the amount of heat necessary to heat the house. Gradually, the water temperature will increase beyond safe limits. A high limit switch is provided to keep this from happening. Even though the room thermostat is calling for heat, the high limit will override that call and shut down the burner. The high limit is nothing more than a thermostat immersed in the

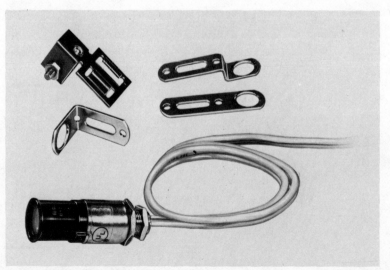

FIGURE 2–7   The cad cell and associated mounting hardware act as a switch in the oil burner circuitry. The electrical resistance of the cadmium sulfide cell decreases when light hits its surface. In the absence of light, resistance remains high and it acts as an open switch. If it fails to "see" light either on startup or during operation, it will shut down the oil burner. (Courtesy of Honeywell Inc.)

FIGURE 2–8   A primary control used in conjunction with the cad cell. Note the transformer at the bottom of the control. The terminal board at the right permits easy connection of the thermostat leads (terminals R and W) and the cad cell leads (terminals F and F). Terminals C, Y, and G are used in heating/cooling applications. (Courtesy of Honeywell Inc.)

boiler water that opens a switch when the water temperature gets too high, say about 200°F.

Now let us suppose that the thermostat has not called for heat for an extended period of time. The water in the boiler might get quite cool and, when the call for heat eventually comes, might take quite a while to warm up to the point where it provides useful heating. A low limit is provided to prevent that problem from occurring. If the temperature falls toward an unacceptably low temperature, the low limit turns on the burner even though the room thermostat does not call for heat. The heat will not get up into the house because the circulator pump, energized by the room thermostat, is still off. The low limit is a thermostat immersed in the boiler water and set to turn on the burner when the temperature gets too low, say about 140°F.

There may be other control combinations, but those described are quite common. Incidentally, because these controls are used on larger systems also, in conjunction with other controls, the reader should keep them in mind.

2.6 SMALL HEATING SYSTEMS—GAS-FIRED

Now let us consider a gas-fired hot air heater, or as it is often called, a warm air furnace.

The room thermostat, on sensing a need for heat, sends a signal directly to the gas valve. Usually, in small systems, the gas valve will be operated by 24 volts and a relay, of the type used in oil-fired systems, will not be necessary. When the gas valve opens, gas flows into the combustion chamber, over the standing pilot, and ignites. At this point we have seen the action of the operating controls, the thermostat and the gas valve.

A major consideration in gas-fired systems, just as in oil-fired systems, is to prove ignition. As mentioned earlier, two techniques are used with gas-fired systems, the standing pilot and spark ignition.

A standing pilot is formed by a small flow of gas, bypassing the main gas valve, going through a pilot safety valve, to a pilot burner. The pilot burner is lit by hand. The flame of the pilot does a number of things. Of course, it is used to ignite the main flow of gas, but it also has a safety function. The flame plays upon a device called a *thermocouple*. This thermocouple is made up of two different materials joined together in what is called a junction. A characteristic of this junction is that it generates a very small

FIGURE 2–9   A pilot burner and thermocouple. The flame impinging on the thermocouple results in a 30-millivolt potential being generated. This is enough to keep the pilot safety switch closed and to permit the main gas valve to function. (Courtesy of Honeywell Inc.)

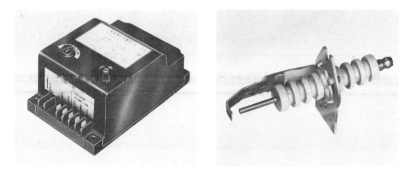

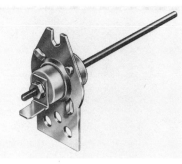

FIGURE 2–10   The solid state module, upon receiving a signal from the thermostat calling for heat, energizes the gas valve and spark igniter. The flame rod must sense a flame within 9 seconds or it will cause the module to shut off the gas valve. The system will attempt to restart twice more. Failing to ignite properly, the system will go into lockout. (Courtesy of Honeywell Inc.)

electrical potential or voltage proportional to the temperature it feels. This voltage has two functions. First, it keeps the pilot safety valve open so that after being lit the pilot flame remains lit. Second, it causes a switch in the electrical circuit to the main gas valve to remain closed so that the valve can respond to signals from the room thermostat.

Proof of ignition is directly related to the presence of a lit pilot. Gas cannot accumulate to an unsafe level in the presence of a flame. If the pilot is lit when the gas flows, there will be ignition. Gas can only flow if the pilot is lit because of the action of the thermocouple. With no pilot there is no voltage generated by the thermocouple and the gas valve will not open even though the room thermostat calls for heat.

Increasingly small gas-fired systems are being provided with spark ignition devices. On a call for heat by the room thermostat, the main gas valve opens and at the same time a transformer creates a high voltage spark across electrodes in the combustion chamber. The gas entering the combustion chamber comes in contact with the spark and ignites. Proof of ignition is usually

Highlights—Gas-Fired Residential Warm Air Heating System

The gas-fired furnace is the most widely used means of heating residences in the United States. The unit shown in Figure 1 is typical of a number of designs on the market. Although the motor shown is of the belt-drive variety, which is quite easy to convert to cooling, the direct drive fan is perhaps even more common.

SEQUENCE OF OPERATION (see Figure 2):

1. On a decrease in temperature in the conditioned space thermostat T will close calling for heat.
2. If the pilot is lit, the pilot safety switch (PS) built into the combination valve CV will be closed. The gas valve GV will be energized with 24 volts and will open. Gas flows through the burner and is ignited by the pilot.
3. As heat builds up and rises above the heat exchanger, the fan switch FS sensing the increase in temperature, will close energizing the fan motor.

(Note the limit switch L in series with the control transformer. In the event that airflow is deficient, due to a dirty filter or inoperative fan motor, the termperature will rise excessively and cause the limit switch to open, shutting down the entire system.)

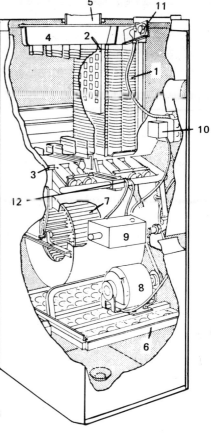

1. Heat exchanger
2. Internal baffle
3. Gas burner
4. Flue—gas collector
5. Flue
6. Air filter
7. Blower
8. Motor
9. Combination valve
10. Transformer
11. Fan—control/high limit
12. Pilot

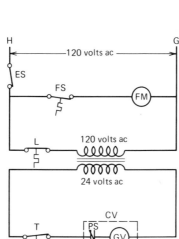

Legend:

FS = Fan switch
FM = Fan motor
L = Limit switch
T = Room thermostat
CV = Combination valve
PS = Pilot safety
GV = Gas valve
ES = Manual "emergency" switch

achieved by means of a flame rod. This device is positioned to sense the flame. If it does so, it sends a signal back to the gas valve indicating that the valve can remain open. If it does not sense a flame within 9 seconds, it will cause the valve to shut off. After several minutes a second effort will be made at ignition and possibly a third. At that point, if ignition is not achieved, the control will "lock-out" and require manual resetting.

An alternate to the flame rod is an ultraviolet sensor that peers into the combustion chamber to see the ultraviolet light emitted by the gas flame. The ultraviolet sensor can see the low level light of the gas flame that a photocell, of the type used in oil-fired equipment, might not detect.

Once ignition has been proved, the heat will build up in the furnace until it trips a temperature switch that will turn on the fan. This causes air to move across the heat exchanger, picking up heat and carrying it to the conditioned space.

Other than the ignition controls the only control commonly found in the hot air furnace will be the high limit. Inadequate airflow across the heat exchanger will result in the air temperature getting uncomfortably warm as it leaves the furnace. Complete failure of the fan motor, a broken drive belt, or some other problem that causes the airflow to stop altogether may result in overheating and cracking of the heat exchanger, or it may possibly lead to a fire. The high limit is intended to prevent this. It is a thermostat mounted in such a way as to sense temperature immediately downstream of the heat exchanger. Sometimes a second high limit, called an auxiliary high limit is mounted very close to the heat exchanger itself. These controls cause the gas valve to close if the temperature rises toward an unsafe level.

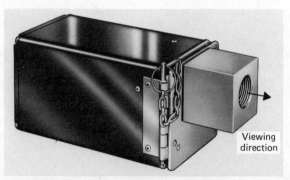

FIGURE 2–11   An ultraviolet detector used on all types of flame systems but particularly on gas-fired equipment, where the intensity of the flame might not be adequate to permit use of the photocell. As with all visual sensing devices, it is important that the viewing element be kept clean. (Courtesy of Honeywell Inc.)

FIGURE 2–12   A control containing a fan switch and a limit switch. As hot air temperature increases, the switch will energize the fan motor. An excessive buildup of heat will cause the limit to shut down the entire system. Note the manual switch that enables the equipment owner to operate the fan constantly if desired. (Courtesy of Honeywell Inc.)

A modification of this control sequence is found in larger gas-fired warm air systems used in commercial rather than domestic applications. The major difference is that the combustion air is provided by a fan instead of by natural convection; it is called a *forced draft furnace*. The room thermostat sensing a need for heat sends a signal to the combustion air fan rather than the gas valve. As the fan operates, pressure is built up in the combustion chamber, which is sensed by a pressure switch. This switch sends a signal to the gas valve when the air pressure has built up sufficiently. On some systems a switch sensing air flow rather than pressure is used for the same purpose. Ignition proceeds from that point as it would in the previously described gas-fired systems.

2.7 LARGER HEATING SYSTEMS

In smaller systems, where one or two operations have to be controlled during an ignition sequence, the controls are fairly simple. In larger systems, however, the complexity increases mainly because the consequences of a failure are increasingly severe. Typically, larger HVAC systems use a steam or hot water boiler ("hot water generator" is a more precise term gaining usage in the field) rather than a hot air furnace. Most of the comments in this section apply to such systems.

More positive control of the ignition sequence is obtained by using programmed flame safety control systems. A timer opening and closing a series of switches will insure that automatic ignition occurs in the proper sequence with proper time intervals between each operation. The controls and their sequence of operation are typically stated in building codes or insurance regulations, which in turn refer to standards put out by Underwriters Laboratories (UL), Factory Mutual (FM), Industrial Risk Insurers [formerly (FIA)], and others.

In order to fully understand a programmed ignition sequence, it is necessary to obtain the manuals and instructions put out by the heating equipment manufacturer *and* the controls manufacturer. Some sequences are quite simple, but others are rather involved. What follows is one such sequence used in starting a hot water boiler firing heavy oil (No. 6 oil). See Figure 2–13.

The thermostat sensing the water temperature determines a need for heat and sends a signal to the programmer. The signal passes through a thermostat that feels the temperature of the oil, which, with No. 6 oil systems, is somewhat critical. If the thermostat senses that the oil temperature is satisfactory, let us say about 225°F, which is adequate for atomizing, it permits the signal to continue on to the programmer timer to start the ignition sequence. If the oil temperature is too low, the sequence would never begin.

The combustion air fan begins to run. The first 60 seconds of operation is called a prepurge period. During this time any un-

FIGURE 2–13   Shown at the left of this mechanical equipment room are two oil-fired hot water generators. Inset A gives a good view of the control panel containing the flame safeguard equipment. Note the wiring diagram and the terminal board, which is useful for troubleshooting. Inset B shows the infrared safety used to view the oil flame, the timer motor used to drive the programmed control, and the switch box containing cam-driven sequencing switches that control the events leading up to safe ignition.

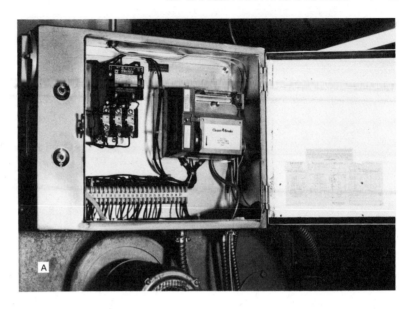

burned combustible gases are forced out of the combustion chamber through the stack and released outdoors.

Within 14 seconds after the fan begins to run, a switch in the airstream of the fan must close, proving that combustion air is flowing. If the switch does not close, the timer will continue to run through the cycle to its starting position and will once again start the sequence. A second failure will result in a "lockout," which

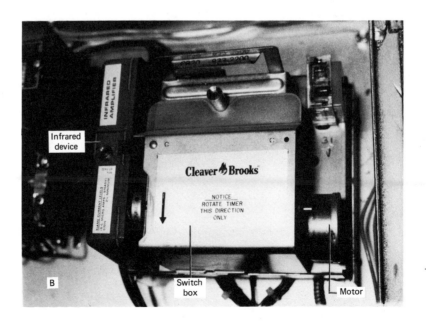

must be manually reset. While the timer is moving forward to its start position, no further switching will occur.

After a total of 53 seconds running time the "low fire end switch," which is mounted on the fuel/combustion air motor, must be in the low fire position. This will ensure that only 60% of the maximum fuel and air enters the combustion chamber at start-up.

At the 60-second point the ignition transformer and pilot gas valve are energized. The electrical spark generated must ignite the pilot before an additional 7 seconds have passed. An ultraviolet scanner views the gas pilot and provides a signal that will enable the sequence to continue. If the pilot fails to light, the timer continues through to its start position and then a lockout occurs. Manual reset will be required.

At the 70-second mark, assuming the pilot has been proved, the main fuel valve opens. At the 80-second mark the transformer and electric spark are shut off. At 85 seconds the fuel/combustion, air modulating motor is switched from "low fire" to "automatic". At this time the damper assumes a position determined by the proportional thermostat, which senses the water temperature. The greater the demand for heat, the wider open will be the combustion air/fuel damper. At 105 seconds the timer stops and the ultraviolet scanner monitors the burner flame as long as the thermostat calls for heat. If, at any time, the flame goes out, the fuel valves will be shut off within 2 seconds and the timer will move back to its start position and lockout.

When the water temperature increases and the thermostat has been satisfied, the fuel valve will shut off, the combustion air/fuel damper will go back to its original "low fire" position, and the timer will restart. It will run for an additional 15 seconds allowing the fan to continue to run through a "postpurge" period removing any unburned combustibles from the combustion chamber.

That completes the combustion control cycle. Other controls dealing with the handling of the hot water produced by the boiler will be dealt with in Chapter 3.

2.8 SENSING THE MEDIUM

In smaller systems the medium being heated is air for forced warm air systems or water for a hydronic heating system. In larger commercial systems air is rarely heated by direct means, hot water and steam being the two main means of extracting heat from the products of combustion.

If an inadequate supply of air passes over the heat exchanger in a warm air system, two things will happen. First, the air itself will become hotter than intended. Excessively hot air flowing into a room can be uncomfortable and can be hazardous if it blows directly on a room occupant. Hot air temperatures should not be much more than 100°F higher than the air entering the furnace. The *limit switch* is intended to sense the temperature of the air leaving the furnace and if it is too high, say above 200°F, it will shut off the main gas valve while leaving the fan blowing.

The second problem is that the furnace heat exchanger may become too hot and crack or the heat may build up to the point where a fire could start. A limit, or an auxiliary limit as it's sometimes called, is mounted so that it can feel the temperature of the heat exchanger even if no air is blowing. It will shut off the gas valve if it senses that an unsafe condition is developing. These same controls are built into oil-fired furnaces as well.

In hot water systems two safety controls commonly sense the temperature of the water. One is a thermostat intended to shut off the burner if the water temperature gets too high, usually a bit below 212°F, the boiling point of water. This would only happen if the normal operating controls malfunction. The second control is a safety valve operated by pressure. The chamber containing the water being heated is called, in addition to a heat exchanger, a pressure vessel. The design of such pressure vessels is prescribed by the American Society of Mechanical Engineers (ASME) Pressure Vessel Code, which states the maximum allowable pressures a boiler shall withstand and also describes the *pressure relief valve* to be used on such equipment. If the pressure builds up because of an operating control malfunction, the relief valve will dump excess pressure so that the heat exchanger does not crack.

Both larger hot water systems and residential systems have these controls. They also may have a control known as a *low water cutoff* that shuts off the burner if there is not enough water in the boiler. If the water level drops too low, the heat exchanger may overheat and be damaged. This device is also commonly found on steam boilers. Hot water boilers are completely full of water while steam boilers are only partially filled, leaving room at the top for steam formation. Because of the great danger of low water level in steam generators, a low water cutoff is found on all steam boilers, even those in smaller applications.

Steam boilers are usually controlled by pressure controls rather than thermostats. The main operating control will be a pressure control, as will the high limit. If the steam pressure starts moving close to an unsafe level, the control will shut down the burner.

A pressure relief valve similar in appearance to the one used on a hot water boiler, but designed for steam use, will dump excess steam into the atmosphere, usually with a loud, startling noise.

FIGURE 2–14 A pressure relief valve that will open at a pressure of 30 psig relieving the pressure within the boiler. Note the capacity of the boiler upon which this device can be mounted (303,000 Btu per hour maximum). Larger capacity valves must be used on larger boilers. Pressure can be manually relieved by lifting the handle at the top of the valve. (Courtesy of ITT Fluid Handling Division.)

FIGURE 2–15   A pressure relief valve mounted on a hot water generator and piped to a floor drain right next to the equipment. Note the insulated hot water supply and return lines. Surprisingly, large amounts of heat can be lost through uninsulated pipes.

2.9 ELECTRIC HEAT CONTROL

Companies selling electric heat have a convincing story to tell. Electric heat is very clean as far as the consumer is concerned, since there is no combustion with the accompanying products of combustion that pollute the atmosphere. Electric heat is quiet, also because there is no combustion process. Electric heat is much easier to control than oil and gas. Furthermore, the initial cost of electric heating equipment is low compared to oil and gas burners. The story has an unhappy ending at this point because only in very few places is the cost of electricity low enough to make operating costs competitive with fossil fuel systems. Where cost is the only factor, electric heat will usually not be used, but where other factors are weighed electric heat has found application and will almost inevitably find further use in the future as fossil fuel costs rise.

Heating electrically is possible because of the fact that an electric current passing through a conductor generates an amount of heat directly related to the resistance of the conductor and voltage of the source of electricity. By using conductors of high resistance, heat can be obtained. Materials such as alloys of nickel and

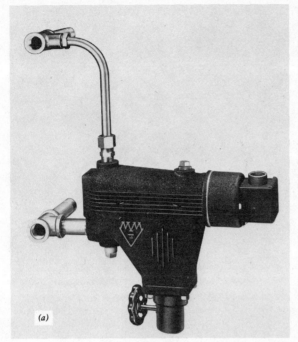

FIGURE 2–16   To prevent damage to boilers in the event of the loss of water, a low water cutoff is used. Two typical such devices are shown here, that at the left used on steam boilers, that at the right used on hot water generators. In both devices a float will open an electrical switch if the water drops to an unsafe level. (Courtesy of ITT Fluid Handling Division.)

chrome, and alloys of iron, chrome, and aluminum, possess characteristics that make them useful in electric heat applications. Such materials are formed into ribbons for use in small radiant heaters; into tubes for baseboard, hot water and duct heaters; and into wire that can be coiled and formed into duct heater elements.

The control of such heater elements can be rather precise, a major advantage of electric heat. Two-position on-off control is most widely used. Thermal lag was previously described as being a shortcoming of such control. When used on electric heating elements, this is not a very severe problem because the mass of the heating element is quite small; it therefore has little thermal inertia. It heats up rapidly and cools off rapidly.

Small electric heaters are often controlled by thermostats directly energizing and deenergizing the heating elements. Baseboard radiators and portable ribbon heaters are usually controlled in this way. As the heater size increases, the need for a relay becomes evident. Thermostats with an ability to carry 20 or more amperes become unwieldy and insensitive to temperature change. The relay will take direction from a small sensitive thermostat and switch on

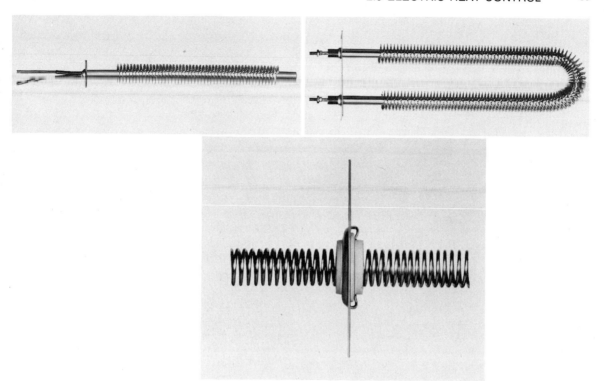

FIGURE 2–17   Two styles of finned tubular element for electric heating units along with a segment of a coiled wire element. All three find application in electric duct heaters, and the finned tubular type is also found in baseboard electric heaters. (Courtesy of Industrial Engineering & Equipment Co.)

very large capacity heaters. The use of such relays enables electric heaters to be energized by stages, a very useful technique.

A two-stage thermostat, upon sensing a requirement for heat, sends a signal to a relay that energizes the first-stage electric heater of, say 5-kilowatts (kW) capacity. If this heating capacity is adequate to heat the space, then the second stage will not be energized. If, however, the temperature continues to drop, the second stage of the thermostat will send a signal to another relay that will energize the second-stage heater of 5 or perhaps 10 kilowatts.

Electromagnetic relays are widely used in larger electric heat applications, where the switching is remote from the conditioned space. Such relays are characterized by clicking sounds as they are energized and deenergized and also by a humming sound, sometimes quite loud. This can be annoying to people in a conditioned space because the electric heaters themselves are essentially noiseless and any extraneous sounds seem even louder. The answer to the problem is found in silent heat-operated relays. On a call for heat by the thermostat, an electric current passes through a small heating element that generates enough heat to cause a bimetal switch to close. A bimetal switch is made of two dissimilar metals bonded together. As it heats up, one metal, expanding faster than the other, causes the switch to bend to a closed position. This is achieved in a matter of seconds, noiselessly.

Two or three increments of heat output can be achieved using staging. The more increments between zero and maximum heat output, the closer the control mode gets to "proportional control." As many as 10 or more stages can be obtained using motorized sequencers. A proportional thermostat sends a signal to a motor connected to a series of switches. As the heating demand increases, the motor rotates and energizes one switch after the other in sequence until the heating demand is matched by the heater output. Although this is not strictly proportional control, it approaches that stage as the number of increments of heat increases.

FIGURE 2–18   Silent operation can be achieved using mercury contactors such as the one depicted. This control is for a three-phase electric heater. (Courtesy of Industrial Engineering and Equipment Co.)

A still closer approach to proportional control is achieved using solid state controllers employing SCRs (silicon control rectifiers). Such devices control either the wave shape of the electrical signal going to the heater or the percentage of time that complete AC waveforms go to the heater. Although the mechanics of what is occurring may seem complex to someone not trained in electronics, the hardware itself is quite simple in application. A "black box" with several wires protruding, to be wired in a particular way, is all that one sees. The future holds more and more solid state devices with higher reliability, simplified maintenance and repair, and very flexible control.

2.10 ELECTRIC HEAT SAFETY

An electric heater can burn out if the proper amount of air does not pass over it. This self-destruction process is a distinct fire hazard. This type of heater requires a safety control system more extensive than most, dictated by UL and the NEC (National Electrical Code) requirements.

Two types of thermal limit control are used: (1) a primary control mounted in the vicinity of the heating elements designed for automatic reset, and (2) a secondary control, which is a backup of the primary, and has a manual reset. The primary control is usually wired in a relay circuit whereas the secondary, set to cut out at a higher temperature than the primary, is usually wired directly to the heating element.

Since the heater will rapidly heat up when signaled to do so, the control system typically has an interlock that ensures that the fan is in operation before the heaters are energized. A sail switch is a device that provides such assurance. It is installed in the duct in such a way as to sense air flowing. The air pushes against a sail that deflects, closing a switch, permitting the heaters to be energized.

Fuses and fusible links are also used to provide safety. Fuses may be located in a control panel, which may be quite close to the heaters in one system or at a remote location in others. They sense electrical current to the heater that could become excessive in the event of a short circuit. The fuse will "blow" or open the circuit if the current exceeds the current-carrying capacity of the wires connected to the heater. Fusible links are usually mounted very close to the heater and respond to either high current, high temperature, or both. The link will melt, breaking the circuit when an unsafe condition exists. When the unsafe condition is corrected, the link must be replaced with a new one before the heater can be put back into operation.

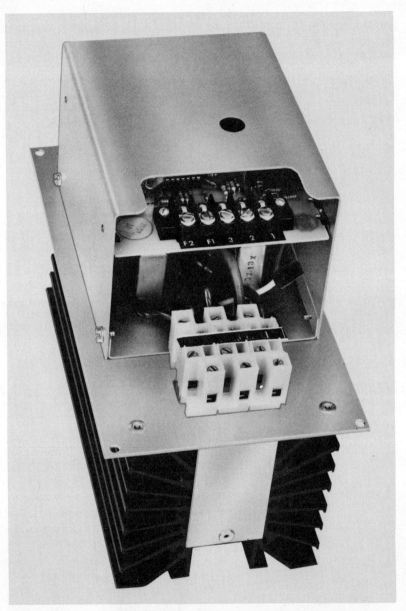

FIGURE 2–19   This solid state controller provides proportional control, silently, by controlling the number of ac waveforms, or cycles that get to the heaters during a specific time period. Most solid state controls are not field-serviceable so that once it has been determined that the device is malfunctioning, it is usually replaced. (Courtesy of Industrial Engineering & Equipment Co.)

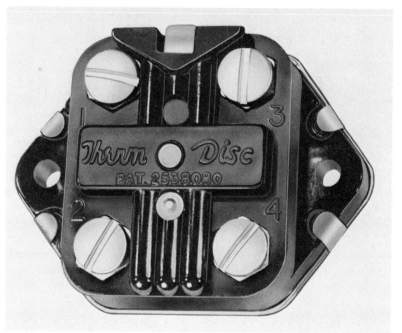

FIGURE 2–20   Two types of automatic reset thermal limit controls. The upper device has an SPST switch and the lower has a DPST switch. Both are bimetal devices and must be in close proximity to the heating element to respond rapidly to increasing temperature. (Industrial Engineering and Equipment Co.)

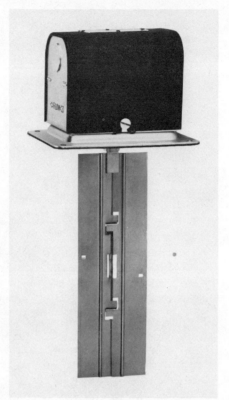

FIGURE 2–21   The "sail switch" will prevent the electric heaters from becoming energized unless an adequate airflow has been established in the duct. Note the direction of the flow arrow on the side of the switch box. Electrically, the flow switch is a single- or double-pole switch that opens at low airflow. (Courtesy of ITT Fluid Handling Division.)

DISCUSSION TOPICS

1. How has the definition of *comfort* changed in the past 200 years?
2. Define *continuous ignition* and *intermittent ignition*.
3. What is staging?
4. Describe the significance of the terms *high fire* and *low fire*.
5. What is meant by *proof of ignition?* How is it accomplished?
6. What is a relay?
7. Describe the similarities and differences between oil and gas ignition systems.
8. How do flame-safeguard systems on larger heating units work?
9. How is the possibility of a boiler failure due to insufficient water in the boiler minimized?
10. What are the benefits and shortcomings of electric heat at this time?

Chapter Three

Distribution of Heat

In the previous chapter we were concerned with how heat was generated. This chapter deals with controlling the movement of heat from the place in which it is generated to the place where it is required, the conditioned space.

Hot water, steam, and hot air are the three main means of moving heat from a central plant to a conditioned space. When hot water or steam delivers the heat to the terminals (radiators, convectors, or fan coil units) in the conditioned area, the system is called a "piped system." When air passes directly over a heat exchanger, where it absorbs heat and carries it through ductwork to a conditioned space, the system is called a "ducted system." Most heating systems in larger commercial buildings are "combination systems" in which steam or hot water carries heat to a heat exchanger or coil in a duct. Air drawn from the conditioned space passes over the heat exchanger extracting heat from the medium and is recirculated back to the space.

In dealing with the distribution of heat, we are interested in how the heat is controlled as it is carried from the central plant and also how it is introduced into the space to provide a comfortable heating effect.

3.1 HOT WATER HEATING

Most hot water systems, both small residential and larger commercial, are low temperature systems (i.e., they operate at temperatures below 250°F); in fact, most operate at temperatures below 200°F. Medium temperature (between 250 and 350°F) and high temperature systems (above 350°F) are usually restricted to large installations in hospitals and industrial plants.

In dealing with an "all water" system, that is, one having no ductwork, the components of the system are the boiler/burner, pump, piping, terminal, and controls. The boiler/burner unit introduces heat to the water, the pump moves the water through the piping to the terminal, where heat is extracted from the water and introduced to the air of the conditioned space. All buildings, large and small have these elements, the major difference being in how they are controlled.

Generally, three techniques of temperature control are applied to hot water systems. First, the temperature of the water in the system may be varied. It seems clear that hot water at 200°F has a greater potential for heating than does water at 140°F. Second, the flow rate of the water through the system may be varied. When the water temperature is held constant, the heat output of the system falls as the flow rate drops and increases as the flow rate increases. A final technique deals with the terminal itself. Varying the quantity of air at the terminal will affect the heat output. With water at a given temperature and flow rate through the terminal, the heat output increases as the airflow increases and decreases as the airflow decreases.

The old cast iron radiator is an example of a terminal, obsolete today, but in widespread use in older buildings. Today's counterpart is the baseboard radiator usually made of copper tubing with aluminum fins for residential use, often found in all steel construction for commercial use. Since air passes over these terminals by natural convection only, the obvious means of controlling heat output would be by water flow or water temperature variation.

The fan coil unit is widely used in hotels, apartments, offices, dormitories, and so forth, to provide individual control of heating. The typical fan coil unit has a metal cabinet, often quite decorative, with a copper tube/aluminum fin coil, a fan, and controls.

FIGURE 3–1   Recessed fan coils of the type shown are used when windows run from floor to ceiling or sliding glass doors occupy the entire wall of a motel or apartment room. The unit at the left is intended to be fully recessed; that at the right is for flush mounting and provides access through the louvered panel. In both systems some ductwork is required to deliver air to the conditioned space. (With permission of the McQuay Group, McQuay-Perfex Inc., HVAC Division.)

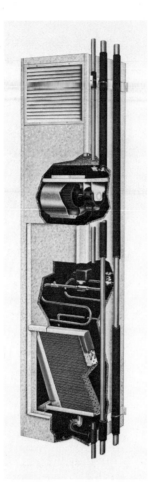

FIGURE 3–2   This style of fan coil unit is popular in high rise applications, where each floor has a similar plan. Units of this type can be stacked one above the other and be connected with vertical runs of piping thereby simplifying installation and reducing costs. (With permission of the McQuay Group, McQuay-Perfex Inc., HVAC Division.)

A variation of the fan coil unit is the unit heater, which is quite utilitarian in design. It incorporates a coil and fan in a housing intended to be suspended from ceilings in factories, warehouses, garages, and similar such work and storage spaces. The controls for the unit heater are usually remote, since the unit itself is installed out of reach.

In small systems the control of water flow is by means of two-position control. The temperature of the water is kept at a fairly constant level by a thermostat built into the boiler. (This was described in Chapter 2.) The thermostat in the conditioned space,

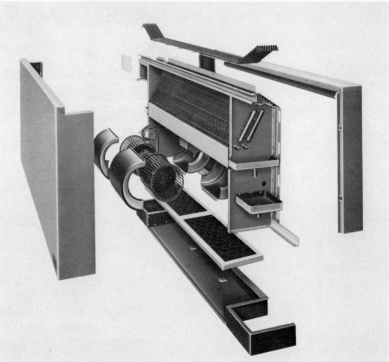

FIGURE 3–3　A decorative fan coil unit commonly found in motels, hotels, dormatories, and the like. It is usually located under the window and may be used for heating only or heating/cooling applications. (With permission of the McQuay Group, McQuay-Perfex Inc., HVAC Division.)

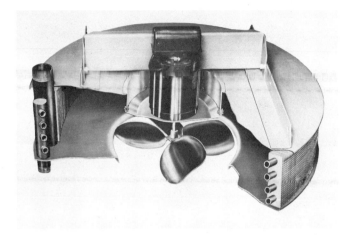

FIGURE 3–4   Two types of unit heater. Both use a fan to blow air into the conditioned space. The unit at the left draws air across the coil and blows it downward; that at the right discharges air horizontally through the coil while discharge louvers direct the air in the desired direction. They are commonly available in steam or hot water configurations. (With permission of the McQuay Group, McQuay-Perfex Inc., HVAC Division.)

when it senses a need for heat, sends a signal to a relay to start the circulator pump. In some commercial applications the thermostat may be of sufficiently heavy duty to be wired directly to the pump; however, the typical small system uses a 24-volt control circuit and requires a relay. This technique can be used in buildings that are zoned also. Each zone, or area, requiring independent control has its own thermostat mounted in the zone and connected to its own circulator pump that supplies hot water only to that zone.

It might seem that having several pumps is an expensive way to provide on-off control to a number of zones. An alternate to that system uses a single circulator and a series of zone valves. Here the zone thermostat controls a valve that will either close or open depending on the requirement for heat in the area. The zone valve used in this way is a *two-way* valve; that is, it has an inlet and an outlet, two ports or connections.

Although some larger systems use on-off control of flow, a more common technique on such systems is to vary the temperature of the water passing through the system in response to either the demand for heat in the conditioned space or in response to chang-

FIGURE 3–5   A booster pump typical of those found in a zoned hot water heating system. Note the arrow showing the direction of flow. (Courtesy of ITT Fluid Handling Division.)

ing outdoor temperature. The device that accomplishes this is a *three-way* valve with three ports or connections.

Understanding the operation of a three-way valve can be a bit tricky. Valve manufacturers identify these valves by the way the water flows through them. If streams of water flow into two of the ports and out the third, or common port as it is called, the valve is said to be a *mixing valve*. If a stream of water enters the common port and is divided into two streams leaving the other two ports, the valve is called a *diverting valve*.

These valves are widely used in large commercial buildings for cooling as well as for heating. They can be used in two-position control as well as proportional control as long as the controller and the controlled device opening and closing the valve are designed for it. They can also be used to control the water temperature or flow rate. These valves will be mentioned throughout the book so that we shall discuss them in some detail.

In Figure 3–8 the hot water flowing into the C port of the valve is permitted to flow either through the fan coil or it bypasses the

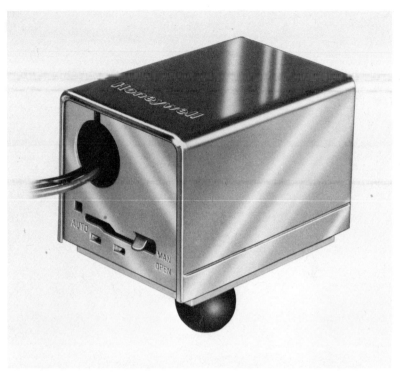

FIGURE 3–6   The operator portion of a two-way zone valve used in small applications. Note that a ball is used to block the flow through the valve. This ball is moved slowly by the motor operator into the open or closed position. The valve shown here has a manual level to permit the valve to be opened in the event of a power failure or if the motor operator becomes inoperative. (Courtesy of Honeywell Inc.)

fan coil, flowing directly back to the return line. The thermostat determines which way the water flows based on the heating demand of the conditioned space. If the thermostat and valve operator are proportional control devices, the valve will move gradually. If there is a great demand for heat, all the water will pass through port B to the coil. As the room heats up, less hot water is required so that more and more water passes through port A and less and less through port B to the fan coil. As the flow rate through the fan coil decreases, the heat output decreases.

In looking at the A and B ports in Figure 3–8, note that they are labeled NC and NO. That designation is commonly used in control diagrams to designate the port as being "normally closed" or "normally open." The term *normally* refers to the valve position when no control signal is being sent to the valve. Port B is a normally open port so that in the event of a power failure, the valve would assume a position permitting flow to the coil. This is a

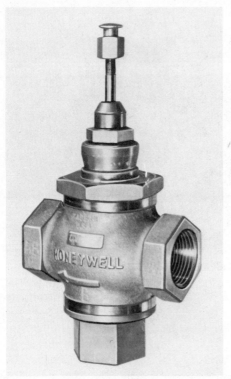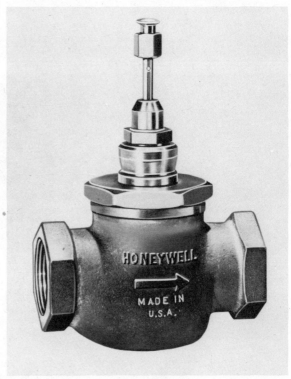

FIGURE 3–7   The direction of flow is important in installing valves. Note the arrow on the three-way valve at the left and on the two-way valve at the right. These valves require an actuator and a linkage before they will operate. (Courtesy of Honeywell Inc.)

common configuration to insure that in midwinter, should the control system break down, hot water would still flow through the heating coil. When a maximum control signal reaches the valve operator, the NC port opens fully and the NO port closes fully. The NC and NO terminology is also used with dampers controlling airflow and with electric switching found in relays and contactors.

In Figure 3–9 control of the heat output is accomplished by controlling the water temperature. A cold water and a hot water line are connected to ports A and B. As the room heating requirement decreases, the thermostat senses it and positions the valve to allow more cold water and less hot water to flow. The flow rate through the coil is constant, but the temperature of the water changes depending upon the ratio of hot water to cold water being mixed at port C.

More involved arrangements using three-way valves will be considered later.

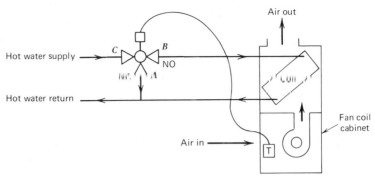

FIGURE 3-8   For clarity the three-way valve is shown outside the fan coil enclosure. In most applications the valve is located within the fan coil cabinet.

Controlling the temperature of water throughout a system, although not usually done in small installations, is very commonly done in larger commercial buildings. Figure 3–10 shows a common arrangement using a three-way diverting valve controlled by a thermostat sensing the outdoor temperature. As the outdoor temperature drops, the valve moves to allow more boiler water to pass through the heat exchanger increasing the temperature of the system water. As the outdoor temperature increases, the valve diverts more hot water back to the return and less to the heat exchanger. The result is that the system water temperature varies with the outdoor temperature.

The use of a heat exchanger here is needed because of boiler performance. On a mild day the required water temperature might be quite low, say 110°F. A boiler attempting to maintain such a low temperature directly would suffer from inefficient combustion and possibly condensation in the flue passages leading to corrosion.

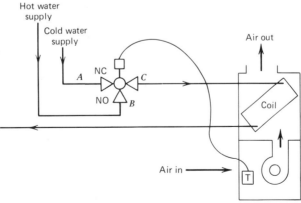

FIGURE 3-9   A hot water supply line connected to the NO port to insure the availability of heat should the control system power supply fail.

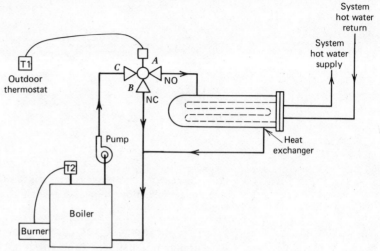

FIGURE 3–10   An example of an *open loop* control system in which the thermostat T1 that controls the temperature of the system water never "feels" the result of its control action. Note that T2 controls the boiler water temperature and is an example of a *closed loop* controller.

Also, some boilers are designed to provide domestic hot water at temperatures of about 140 to 160°F and would of course not be able to do so if they were required to drop down to 110°F. In the system shown, the water is maintained at about 200°F. This hot water is passed through the heat exchanger in varying quantities, dependent upon outdoor conditions. It never mixes with system water, merely heats it. This system water then flows to baseboard radiators, fan

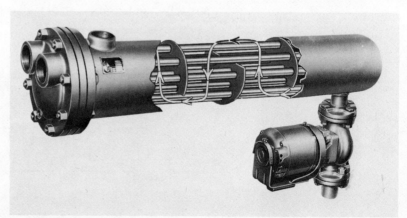

FIGURE 3–11   A water-to-water heat exchanger in which the boiler water is introduced at the left, flows through the small diameter tubes, and leaves at the left. The system water is circulated around the tubes by the water pump and leaves at the top. (Courtesy of ITT Fluid Handling Division.)

coils, unit heaters, and so forth, at a temperature warm enough to do the job, yet cool enough to minimize thermal overshoot and piping losses.

Note in Figure 3–12 that two thermostats are used: T1 senses the outdoor air temperature and T2 senses the temperature of the hot water being supplied to the building. This is called a "master-submaster" arrangement with T1 being the master thermostat and T2 the submaster. The temperature of the supply water is controlled by T2 causing the three-way valve to allow more or less boiler water to enter the heat exchanger. The temperature or setpoint that T2 will hold is dictated by T1 based on outdoor temperature. As an example, if the outdoor temperature were 0°F, the setpoint of T2 might be 200°F. As the outdoor temperature increases to 20°F, the water temperature might only have to be 180°F to adequately heat the building; at a 40°F outdoor temperature perhaps 160°F water would be adequate. For every degree of increase in the outdoor temperature, as sensed by T1, the setpoint of T2 decreased 1°F. This relationship is called the *reset ratio*. In the example given, the reset ratio is expressed as 1:1 (one to one). In a system with a 1.5:1 reset ratio, a 1½ degree change in the outdoor temperature will produce a 1 degree change in the set point. In some controls the reset ratio is fixed at the factory whereas others provide the flexibility of being able to change the reset ratio in the field to match specific building and climatic conditions.

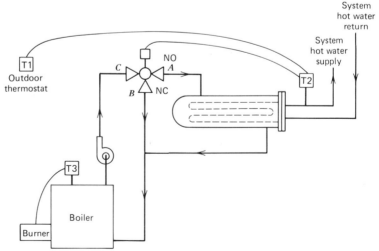

FIGURE 3–12   Master thermostat T1 signals the submaster thermostat T2 directing it to maintain a particular system water temperature based on the outdoor temperature. Thermostat T3 is completely independent of T1 and T2, maintaining boiler temperature regardless of outdoor conditions.

3.2 STEAM HEATING

In hot water heating we saw that water was used to extract heat from the products of combustion. In steam heating systems the same process is used; the difference is that the water is allowed to boil off into vapor or steam. The pressure generated by this process forces the steam to remote terminals, where the steam condenses as it gives off heat to the conditioned space. The elements of a steam heating system are the boiler/burner unit, piping, terminals, and controls. Although space heating systems are still being designed and installed using steam, the trend has been toward hot water systems in recent years.

Hot water systems use three control techniques, but steam systems typically use only two. The temperature of the steam is held constant by the boiler controls. Control of the heat output is obtained at the terminals by means of two-way valves, either two position (on-off) or proportional, designed for steam use. Control of the on-off action of a fan blowing air across a coil, the fan coil arrangement previously discussed, is also used.

Of perhaps more importance than all-steam systems is the combination steam–hot water system. This is a popular heating system in locations where steam is already available, such as in modernizing or expanding existing facilities or using waste steam produced in an industrial process for space heating. The steam is passed through a heat exchanger to heat water and the water is then pumped to terminals in the manner previously described. The heat exchanger, sometimes called a *converter,* substitutes for a hot water boiler in this arrangement (see Figure 3–13).

Figure 3–14 shows a system employing a converter. Thermostat T senses the supply water temperature and signals the steam valve to either increase or decrease the steam flow rate to maintain the water temperature. Thermostat T could be reset by an outdoor

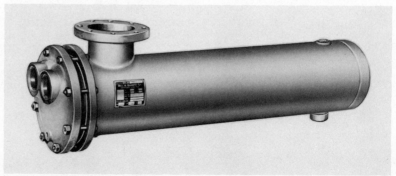

FIGURE 3–13   A converter similar in appearance to the water-to-water heat exchanger. Water flows through the left end of the converter while steam is introduced at the top left and condensate leaves from the port at the bottom right. (Courtesy of ITT Fluid Handling Division.)

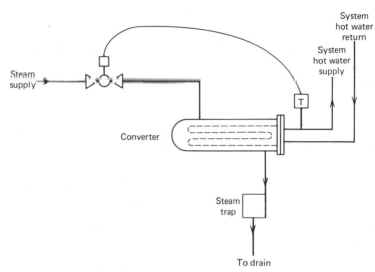

FIGURE 3–14   A steam converter useful in areas where a cheap supply of steam is
available. Steam control valves are invariably two-way valves. Note the
steam trap location. Its function is to hold back steam in the converter and
ensure that only water flows to the drain.

thermostat or by a thermostat located in the building although that
is not shown here. The steam enters the converter where it condenses, giving up heat to the water. The condensate flows out the
bottom of the converter to a drain or perhaps is recirculated back
to the plant steam boiler. The purpose of the trap shown is to
insure that only condensate leaves the converter and not steam.
Any steam leaking through is lost energy and the equivalent of
pouring money down the drain.

3.3 HOT AIR HEATING

Direct-fired hot air systems, that is, systems in which the air
diffusers, wall registers, or troffers positioned in such a way as to
tion are flowing, are used in residential and smaller commercial
installations. Although some larger commercial installations may
use direct-fired furnaces, the application is one in which many
smaller units are employed to do a big heating job.

The control of such units is generally two-position or staging as
explained in Chapter 2. The air is carried through the duct system
to the conditioned space, where it is introduced through ceiling
diffusers, wall registers, or troffers positioned in such a way as to
provide even air distribution.

At different times of the day different areas of a home or office
building may have different heating requirements. The solution to
the problem of providing varying amounts of heat to different spaces
using a single furnace is called *zoning*. In hot water systems we
saw that zoning could be achieved by using several circulators or

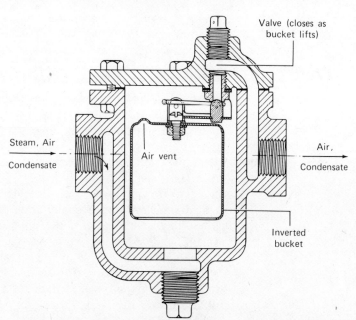

FIGURE 3–15   An inverted bucket trap commonly used on steam converters. Condensate fills the trap and overflows to the drain. If steam or air enters the trap, the bucket lifts sealing the valve, thus preventing steam from being lost. The air passes through the air vent and is passed into the drain when the condensate again builds up to a level sufficient to cause the valve to open. (Courtesy of ITT Fluid Handling Division.)

zone valves. In hot air systems the equivalent of a zone valve is used, namely a zone damper. The damper is a device that controls the flow of air through a duct upon command of a zone thermostat.

In the system shown in Figure 3–16 the controls are designed in such a way that if any zone thermostat calls for heat the zone damper opens and the furnace comes on. The furnace will continue to run as long as any zone calls for heat. When the last zone damper closes, the furnace shuts down. This action is usually obtained by having electric switches called "end switches" or "auxiliary switches" built into the motors driving the dampers. The switches make or break the electric circuit to the furnace when the damper has moved to a fully open or fully closed position.

A real problem in this type of system, which is called a variable air volume system, is that as the dampers close, air pressure builds up and airflow through the furnace decreases. Controls must be used to dump the excess air pressure to the return side of the furnace. This keeps the duct pressures under control along with accompanying air noises and air discharge patterns in the room. Unfortunately, it builds up the temperature of the air entering the furnace, which if allowed to rise unchecked, will trip the high limit safety. The solution to this problem is either a furnace with staging or flame modulation, since most of these furnaces are gas-fired.

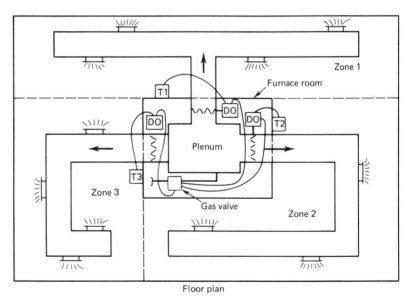

Floor plan

FIGURE 3–16   Zones 1, 2, and 3 are conditioned by air controlled by damper operators that respond to temperature changes as sensed by zone thermostats T1, T2, and T3. An end switch on each damper operator is connected to the gas valve. When the last zone requiring heat has been satisfied, the gas valve is shut off.

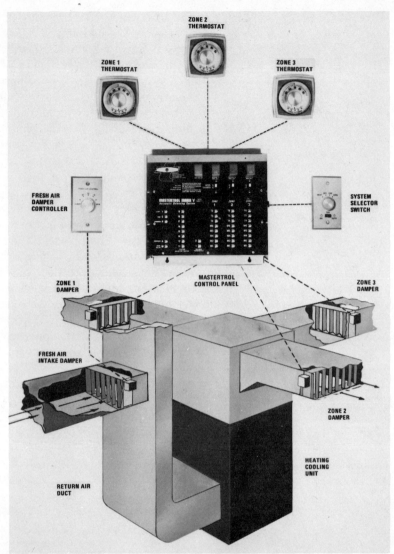

FIGURE 3–17   A variable air volume system using duct dampers controlled by zone thermostats through a master control panel. Typically, 24-volt control systems are used. (Courtesy of Trol-A-Temp, Division of the Trolex Corp.)

Control of the heat output would be accomplished with a thermostat located in the discharge duct that cuts back on gas flow as the temperature rises. If the source of heat is electric heaters, control becomes rather easy by staging or other techniques described previously.

The control of "heating only" systems is quite simple in most cases because zoning usually is not required. Where it is required, the controls can be quite complex and expensive. Such an expense is not warranted in most heating only applications. Zoning is quite popular in larger systems that employ cooling as well as heating. Such installations require multizone systems, which will be described in detail later.

3.4 COMBINATION HEATING SYSTEMS

In large installations it is quite common to see a boiler in a central location producing hot water that is pumped to coils mounted in ducts in one or more remote locations. Such a system, which employs piping and ducting to get the heat from the central plant to the conditioned space, is called a "combination system."

The quantity of heat being added to a conditioned space can be controlled in two ways: (1) by controlling the temperature of the air, or (2) by adjusting the volume of air entering the space. Both of these techniques are used, but *variable volume–constant temperature* systems are generally used in heating/cooling applications whereas *constant volume–variable temperature* systems are used in heating only or cooling only applications.

The control of the temperature of the air in a duct is obtained in several ways. One is to control the volume of water passing through the heating coil. At low flow rates the velocity of the water is lower and more heat is extracted from each pound of water passing through the coil. Fewer pounds of water are circulating, however, and the net effect is lower heat transfer. This can be readily seen in Figure 3–18. In a coil with high water velocity the average water temperature is 80°F higher than the air passing over the coil. In the coil with low water velocity the average temperature of the water is only 50°F higher than the air. It is apparent that greater heat transfer will occur in the case of higher water velocity and therefore higher water flow rate.

The flow rate through a duct coil can be controlled using three-way mixing and diverting valves. Figure 3–8 showed how a divert-

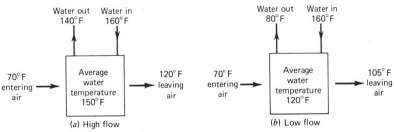

FIGURE 3–18 · The high water velocity in the coil shown in *(a)* results in a higher coil temperature and therefore greater heat transfer than the low flow situation shown in *(b)*.

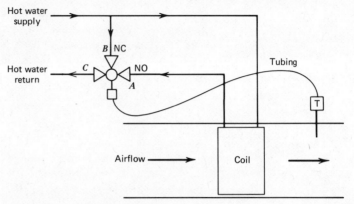

FIGURE 3–19   This three-way valve is actually a mixing valve, since it has two streams entering the valve and one leaving. The application is as a diverting valve because more or less water flows through the coil depending on the position of the valve in response to thermostat T.

ing valve could be used on small fan coil units. The same technique using larger valves to accommodate larger water flow can be used on duct coils. The use of a mixing valve is shown in Figure 3–19. In this case the actual application of the valve differs from the manufacturer's designation of that valve so that things get a bit difficult. In this figure we have a *mixing valve* used in a *diverting* application. The thermostat downstream of the coil causes the valve to shift in such a way as to cause more or less water flow through the coil. The flow rate from port *C* is constant but as the heating demand increases, as sensed by thermostat T, the valve shifts to allow port *A* to open more and port *B* to close down a bit. The heat output of the coil then increases. The three-way valve shown here could also be a two-position valve that allows either full flow or zero flow through the coil, or it could employ a floating control with a deadband.

The control of coil temperature has also been described for small fan coils using a three-way mixing valve. Similar control for duct coils can be used when a source of cold water is available to mix with the hot water. In the absence of such cold water another technique can be used as depicted in Figure 3–21. The common port of the three-way mixing valve leads directly to a pump. This pump forces full flow through the coil constantly. The three-way valve is controlled by thermostat T, which opens port *A* fully on a demand for maximum heat. As the demand drops, port *A* closes and port *B* opens. Since the valve is on the intake side of the pump, the flow of water is from the return line *into* the valve through port *B*. Water from the return line, which has cooled down since it has already passed through the coil, is recirculated through the coil. The water temperature entering the coil is lower because

FIGURE 3–20   This pneumatically operated three-way valve mixes the return and supply water to control temperature. Note the manual two-way valves installed to permit the pneumatically operated valve to be removed for service without having to drain the entire system.

hot water from the supply line and cooler water from the return line mix. When there is no demand for heat the valve moves to close off port A totally and open port B fully so that flow is still constant through the coil but the water temperature is very low. In that case there is no water coming from the supply or leaving through the return line; full flow is just being pumped around a loop. This would not work without the pump, since the pressure at C must be lower than the pressure at both A and B. This can only occur when the pump is in operation.

In the diagram shown, T is sensing the duct temperature and controlling the valve in response to temperature changes. It could just as well have been located in the conditioned space. However, the duct location results in a faster response to temperature changes. Another possibility would be to have T act as a submaster by having its setpoint *reset* by an outdoor thermostat or even by a thermostat located in the conditioned space. Still another configuration has T used as a high limit designed to override the direction from the main thermostat in a conditioned space. The main thermostat might demand that the water valve shift to admit more hot water to the coil, but the high limit thermostat senses that the air temperature is too high already and prevents the valve from responding to the main thermostat, and might even cause the valve to reduce the hot water flow somewhat. Such an arrangement can

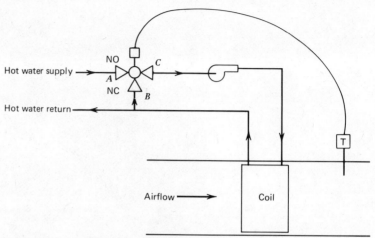

Hot water supply →

Hot water return →

NO

NC

A

C

B

T

Airflow →   Coil

FIGURE 3–21   A three-way valve that acts as a mixing valve, since it is mixing hot supply water with cooler return water. If port *A* were totally closed, the flow from port *C* would still be constant; however, the water temperature would soon approach the temperature of the air in the duct system with no heat being added to the air.

reduce the temperature overshoot and also prevent uncomfortably hot air from being discharged into the rooms.

In studying the various diagrams of coils, note that they all have something in common. In each case the supply water enters the coil on the side that the air leaves. The water flows through the coil in one direction, the air in the opposite direction. This is true of all heat exchangers whether they are air to air, air to water, or water to water. This is called "counterflow" and is necessary for maximum heat transfer efficiency. All heat exchangers and coils should have their inlet and outlet marked as such, but should this marking be missing, remember that counterflow is mandatory.

In discussing combination systems, we have been concerned with what is happening inside the duct coil. However, we can also control the air temperature by manipulating the air in the duct rather than the water in the coil. A popular control technique that employs face and bypass dampers is shown in Figure 3–22. The face dampers are those shown controlling the air passing through the coil. They cover the "face" of the coil, hence the name *face dampers*. The bypass dampers control the airflow moving around the coil or "bypassing" the coil. The dampers are controlled by a damper operator, which causes the face dampers to close and the bypass dampers to open on a signal from thermostat T. This arrangement can be compared to the three-way mixing valve in which two streams of water of different temperatures are mixed to provide a third temperature. In the case shown here untempered air through the bypass dampers and tempered air through the coil are

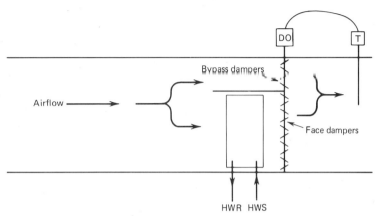

FIGURE 3–22   In systems employing face and bypass dampers, there is no need for control of the water temperature or flow. Hot water flow is constant as long as the boiler and main circulator are in operation. Temperature control is achieved as the damper operator responds to thermostat T.

mixed downstream to provide the air temperature necessary to satisfy the requirements of the conditioned space.

The bypass dampers are smaller in size than the face dampers because of the restriction offered by the duct coil. If no heat is required, all the air may pass through the bypass, whereas if maximum heat is required, all air must pass through the coil. Since there is no restriction in the bypass duct, it can be made small compared to the duct that has the highly restrictive heating coil. In a good design the restriction to airflow offered by both the bypass with its dampers and the heating duct with its coil and dampers is equal.

Control of the damper operator is similar to control of the three-way mixing valve. Thermostat T can be located in the conditioned space, can be a *submaster* responding to an outdoor or conditioned space master, or it can be a high limit.

3.5 PREHEAT AND REHEAT

The main goal of a heating system is to add heat to a building at a rate that matches the rate at which it is being lost. In large installations the main heating coil is designed to do this. There are secondary goals that the HVAC system design strives to reach that also require heating; as we get more involved in these larger systems, the terms *preheat* and *reheat* will be used.

It is the rare HVAC system these days that calls for heating only. Most systems require outside air to be brought into the building for ventilation and, in some cases, to produce a cooling effect. Dry air at a temperature below 65°F can do this rather well. Even on the coldest days there is a requirement for ventilation air and conceivably for cooling air in certain parts of a building. It is undesirable to have air entering an airhandler at a temperature much below 55°F. We shall consider this requirement in greater detail later, but for now we just assume that air at a lower temperature produces uncomfortable drafts, and if lower than 32°F, can lead to water freezing in pipes. Freezing water expands, pipes rupture, and repair bills soar—a very unhappy state of affairs.

A coil located in a duct close to the outdoor air intake, designed to raise that outdoor air temperature to about 55°F before it gets to the main airhandler, is called a *preheat coil*. The heat output of the coil is usually regulated by regulating the water flow by means of a two-way valve or three-way valve controlled by a thermostat sensing the air temperature on the leaving side of the coil. As the temperature drops below 55°F, the thermostat directs the valve to increase the flow to the coil heating the air.

In many cases the valve is designed in such a way that even if the control system fails, perhaps because of a power outage, the valve will not close entirely. As long as the water is in motion, even if it is not too warm, the chances of a freezeup are quite small. If the water were to stop altogether, it would freeze quite rapidly at air temperatures at or slightly below 32°F, while even slow-moving water could remain liquid at air temperatures as low as 0°F. An added safety precaution is the use of a thermostat connected to the air handler fan motor. It senses the air entering the air handler and if it is too low, say below 45°F, it shuts down the fan motor. These controls are very common even in applications where preheat coils are not used.

The reheat coil, as the name should indicate, heats the air again, or reheats it, after it passes through the main HVAC unit. These reheat units are usually rather small, generally being required to add only a small amount of heat to the air. One common requirement for reheat is on systems that provide dehumidification using refrigeration. The air passes over a cold coil that removes water from it, the air which is cooled in the process passes over a reheat coil where its temperature is brought up to a comfortable level. Another use for reheat coils is in HVAC systems that provide the same temperature air to all parts of a building, even though the heating requirement may be different. A reheat coil, controlled by a thermostat in the individual space can control the temperature of the space by adding small amounts of heat on demand.

The control of reheat is the same as control of any hot water or steam coil as previously described. The thermostat controlling the reheat function can be located in the duct on the leaving side of the coil or in the conditioned space.

Although reheat and preheat have been presented in terms of hot water coils located in a duct, other heating means are possible.

Electric heat is quite commonly used in reheat applications. Oil-fired and gas-fired duct heaters can also be used but are not nearly as convenient or as common as the other means described.

DISCUSSION TOPICS

1. How may the heat output of a terminal unit be controlled?
2. Explain the difference between a diverting valve and a mixing valve.
3. What do the symbols NO and NC mean?
4. Explain how a master–submaster control arrangement works. What is a reset ratio?
5. How can a converter be used in a heating system?
6. What is zoning?
7. How can a mixing valve be used to control the coil temperature? A diverting valve?
8. What is the difference between preheat and reheat?
9. Explain counterflow.

Chapter Four Producing a Cooling Effect

In Chapter 2 we described ways in which heat was generated in a central plant for distribution throughout a building. The heating equipment was sized to add heat to the building at a rate that matched the rate at which it was being lost to the outdoors. In this chapter we shall take a similar approach except that now we are concerned with equipment that removes heat from the building at a rate equal to the rate at which it enters due to warm outdoor air temperatures, infiltration, solar effect, and internal load.

The equipment with which we are concerned transfers heat from the building to the outdoors by mechanical refrigeration. Although other systems, such as absorption systems, are also in use, by far the greatest number of buildings from small homes to large commercial structures are cooled by mechanical systems using reciprocating, centrifugal, or screw compressors.

4.1 SMALL DIRECT EXPANSION SYSTEMS

In direct expansion refrigeration systems the condensing unit produces a liquid refrigerant that is controlled by the refrigerant flow control as it passes into an evaporator coil (DX coil) located in a duct. Such an arrangement is found in small residential and commercial applications and even in larger applications where fairly small units, say up to 50 tons of refrigeration, are used in multiples.

In a typical small installation a room thermostat designed for on-off two-position control sends a signal to a relay. The relay in turn energizes the compressor, condenser fan, and air handler or evaporator blower. The system operates, sending cold air to the conditioned space until the thermostat senses that the temperature is sufficiently low and shuts the system down. Incidentally, timed two-position control can be used in this application. The thermostat must be equipped with a "cooling anticipator," a small electric heater that generates heat only when the thermostat is satisfied. This small amount of heat generated within the thermostat provides

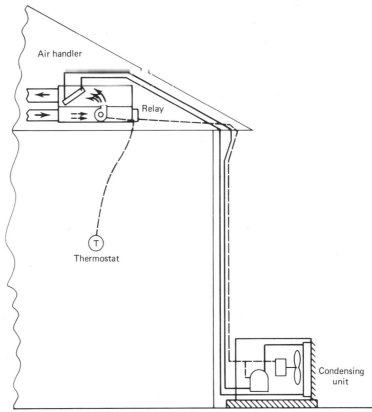

FIGURE 4-1   In this control arrangement a wall-mounted thermostat sends a 24-volt signal to the relay that controls the air handler and the condensing unit. The transformer may be located in the condensing unit or in the air handler. In some installations it is an accessory item purchased separately.

a false signal to the thermostat and causes it to turn on the cooling system a bit early, reducing the operating differential in much the same way as the heat anticipator discussed in Chapter 1.

Other than an on-off thermostat these smaller systems have only enough controls to protect the equipment from self-destruction, much as the heating equipment mentioned earlier had. Abnormally high or low refrigerant pressures can cause inefficient operation and premature equipment failure so that pressure controls are installed, which are designed to shut the equipment down if unsafe conditions occur. There are two schools of thought on whether these controls should be reset manually or automatically. Those who prefer manual reset devices claim that the equipment owner/operator is signaled that an impending major failure is likely and is made aware of this by the necessity of resetting the control. Those favoring automatic reset controls claim that nuisance trips or

Highlights—Single-Package Heating/Cooling Unit

The single-package air conditioner, the wiring diagram of which is shown in Figure 1, has a 5-ton 240-volt single-phase compressor and two-stage electric resistance heat.

SEQUENCE OF OPERATION

Cooling Mode

1. On a call for cooling the thermostat makes the 24-volt circuit from R to G and Y.

2. *Relay coil RM* is energized causing the *RM* switch between 4 and 2 to close. This energizes the indoor blower motor. (Note that this motor has four speeds. Only the HI speed is wired in this unit. If lower airflow is desired, the motor must be rewired.)

3. Contactor coil *CC* is energized causing the switches labeled *CC* to close, energizing the compressor and outdoor fan motor.

Heating Mode

1. On a call for heat the thermostat makes the 24-volt circuit from R to W1.

2. Heat operated relays RH1 and RH2 are energized. The RH1 switches close energizing the indoor fan motor and heater HTR1. The RH2 switch closes energizing relays RH3 and RH4. The RH3 switches close energizing heater HTR2 and the indoor fan motor. (Note: Since the RH1 switch has already energized this motor, the action of RH3 has no effect. It is intended as a back-up in the event of failure of the RH1 switch.) The RH4 switch closes.

3. If the conditioned space temperature continues to fall, the thermostat will make the 24-volt circuit from R to W2.

4. Since switch RH4 is already closed (note that an interlock with the first stage of heating is thereby provided), the RH5 relay will be energized. The RH5 switches close energizing the fan motor (a redundant safety) and heater HTR3.

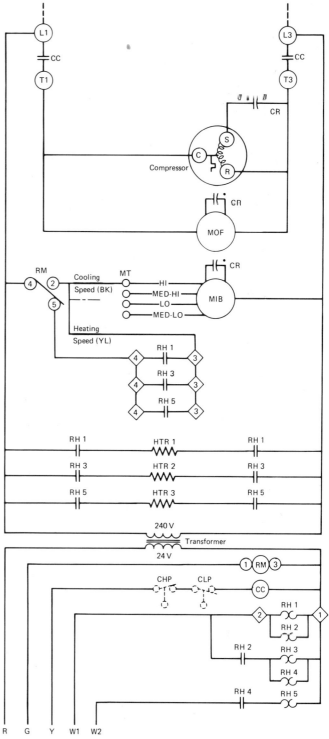

Legend:

CC	=	Contactor
CHP	=	Control, high pressure
CLP	=	Control, low pressure
COMP	=	Compressor
CR	=	Capacitor, run
CS	=	Capacitor, start
MIB	=	Motor, indoor blower
MOF	=	Motor, outdoor fan
MT	=	Motor terminal
RH	=	Relay, heater
RM	=	Relay, blower motor
RS	=	Relay, start
TRANS	=	Transformer
HTR	=	Heater, electrical resistance

shutdowns causing excessive manual resets motivate an equipment operator/owner to bypass or mechanically deactivate the device leaving the equipment with no protection at all. They claim that at least with an automatic reset device the equipment is always protected and an eventual reduction in performance will signal the owner/operator that service is required. In any case, *both* types of control are in widespread use.

In addition to protecting the refrigeration system, controls are provided to monitor the current draw of the compressor motor and occasionally of the fan motors as well. Such devices sense the overcurrent caused by voltage problems or by excessive loading of the motors. They are typically automatic reset devices although in some instances a manual reset is used.

4.2 LARGER DIRECT EXPANSION SYSTEMS

Control systems on larger direct expansion installations are much more interesting and varied than those on smaller residential equipment. One reason is that such smaller systems are at the low end of the price scale and are therefore made as inexpensively as possible to compete in the mass consumer market. Although deluxe units are available incorporating better control systems, they are in the minority.

In larger direct expansion systems the consequences of a system's being out of service are usually greater and a customer is quite often willing to pay extra for equipment that is more reliable. An oil pressure switch is one such piece of equipment found on larger systems. It senses the adequacy of the lubricating oil pressure in the system, and if such pressure is inadequate it will shut down the compressor to prevent major failure. The typical oil pressure switch has a built-in time delay that will permit the compressor to run for a minute or a bit longer while the oil pressure is low. If the pressure does not build up after that time, the switch shuts off the unit. Since most such switches are manual reset types, this time delay eliminates a large number of nuisance service calls that could be caused by random short-duration oil pressure shortages.

The crankcase heater is a device intended to keep the lubricating oil in a compressor warm so that it flows in cold weather even after the compressor has been idle for an extended period. It also prevents the refrigerant from mixing with the oil and diluting it, reducing its lubricating ability. In some systems the control arrangement has the crankcase heater energized only when the compressor is off, whereas in others the crankcase heater is always hot.

Another control commonly found in larger systems is an antishort cycle timer, which comes in a variety of shapes from a simple clock motor arrangement to a "black-box" solid state device. Its intent, however, is quite simple, to prevent the compressor from cycling rapidly, either in response to erratic control operation, or if automatic reset pressure or current overload devices are used. Such rapid cycling could lead to early failure of the compressor. Generally, the operating characteristics of these devices are similar. When the compressor stops, due to normal thermostat operation or in the event of a safety control's responding to an unsafe condition, a period of 5 to 10 minutes must pass after the thermostat once again calls (or the safety switch resets) before the compressor will attempt to restart.

Other controls found in larger direct expansion systems are there for two main reasons. First, the equipment in larger buildings is called upon to operate for as long as 8 or 9 months of the year, in some cases year-round, as compared to the typical small residential cooling system that may only operate from mid-May to late September. In order to operate reasonably well when the outdoor temperatures are low, control of the high side pressure must be maintained.

Second, the cooling load of the building may vary considerably from week to week, and even during the course of the day as the sun moves across the sky. The cooling system should have something more sophisticated than two-position on-off control. This requirement for better control can be met with direct expansion systems but not as easily or as accurately as with chilled water systems, which will be discussed later.

The control of high side pressure is achieved in several ways. Perhaps the most common is to have a number of condenser fans cycle on and off in response to the outdoor temperature or alternatively, in response to high side pressure. One such system has three fans; one runs continuously; one runs only at outdoor temperatures above 75°F; and the third runs only at outdoor temperatures above 85°F. This might be recognized as an open loop system. We are actually trying to control high side pressure but the controller, the outdoor thermostat, is not sensing this pressure. The use of a reverse-acting pressure control would seem to be a better technique. The fans could be cycled on at a pressure rising toward 250 psig (assuming an R-22 system) and shut down as the pressure falls to 200 psig. Sequential operation could be obtained by using several pressure controls, each with a slightly higher cut-in pressure.

Cycling the fans is one way of controlling the airflow over the condenser coil and thereby the head pressure. Dampers over the face of the condenser is another way of doing the same thing. The damper operator can be controlled by high side refrigerant pressure directly. At high condensing temperatures and pressures the dampers are wide open allowing maximum airflow while at lower temperatures and pressures the dampers close down. Such systems can work well even when the outdoor temperatures drop below freezing.

Another method of high side pressure control uses a regulating valve that is located between the condenser and receiver in the refrigeration system. As the high side pressure drops, the valve closes, holding back the refrigerant and flooding the condenser. As more and more liquid refrigerant accumulates in the condenser, less and less surface is available for condensing, with the result that the high side pressure increases. Although such systems typically have large charges of refrigerant, a good degree of control is attainable.

Having assured that the cooling equipment will work year round if need be, we now discuss some common techniques used in regulating the cooling capacity of direct expansion equipment. Two position on-off control is used in some smaller commercial equipment. The manner in which the equipment is shut off and started up is called a ''pump-down'' cycle. In this system a thermostat in the conditioned space becomes satisfied and sends a signal to a solenoid valve in the refrigerant liquid line causing it to close. The compressor continues to run, pumping refrigerant from the low side of the system into the high side, where it is stored in the receiver. In some air conditioning systems having no receiver the condenser is sized large enough to hold the entire refrigerant charge. Eventually, the low side pressure drops, after most of the refrigerant has

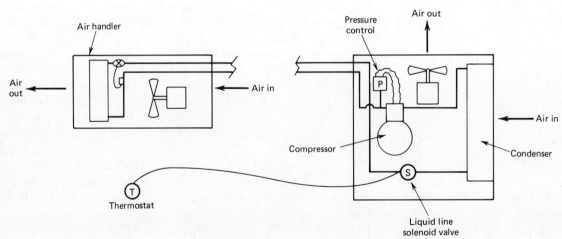

FIGURE 4–2 A pump-down system in which the condenser is oversized to accept the total refrigerant charge. In some systems a receiver is added for this purpose. Note that the thermostat controls the liquid line solenoid valve and the pressure control turns the compressor on and off.

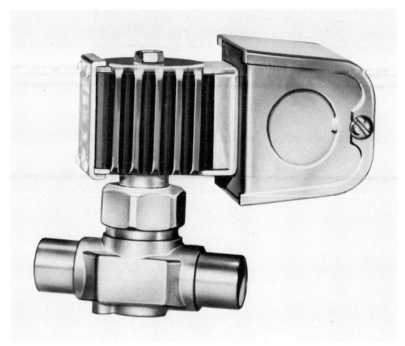

FIGURE 4–3   A solenoid valve typical of those used in pump-down applications and also in pneumatic control systems. The encapsulated coil is replaceable and is available in a variety of voltages. (Courtesy of the Sporlan Valve Company.)

been pumped into the receiver, and a pressure control sensing this pressure deenergizes a relay, which in turn shuts off the compressor. On a call for cooling, the reverse sequence occurs with the thermostat opening the solenoid valve. This allows the refrigerant to flow to the low side of the system increasing the pressure to activate the pressure switch and turn the compressor back on. Pump-down cycles are designed to protect the compressor by preventing the liquid refrigerant from migrating to the compressor crankcase during the off cycle and mixing with and diluting the oil. It also permits the compressor to start under a low load condition, which decreases wear and tear on the motor.

Capacity control that relies strictly on turning a compressor on and off is pretty much restricted to smaller residential applications and light commercial systems. The next step up in sophistication is the staging of compressors. A two-stage thermostat can control two compressors sequentially as the cooling load requires. In some cases four compressors might be controlled in pairs by such a thermostat. The object of such control is to prevent all compressors from cycling on and off together. The power inrush when large capacity compressors are energized can be quite high.

Highlights—Control Settings

Controls may often be adjusted in the field to provide desired operating characteristics. Three control settings of importance are related by the following equations:

    a. For controls that open on a rise in temperature/pressure:

$$\text{differential} = \mathit{cut\text{-}out} - \mathit{cut\text{-}in}$$

    b. For controls that close on a rise in temperature/pressure:

$$\text{differential} = \mathit{cut\text{-}in} - \mathit{cut\text{-}out}$$

If we know two of the values in these equations, the third can be determined. Field-adjustable two-position temperature and pressure controls, such as those shown, must be accurately set to provide the desired action.

The temperature control of Figure 1 has a cut-in temperature of 50°F. Note that the control is designed to open, or cut the equipment out, on a rise in temperature. The differential is about 7°F. Therefore, the cut-out temperature will be:

$$
\begin{aligned}
\text{cut-out} &= \text{differential} + \text{cut-in} \qquad \text{(from part (a) above)} \\
&= 7 + 50 \\
&= 57°F
\end{aligned}
$$

The right side of the pressure control in Figure 2, which is designed to close on a pressure rise, has a cut-in pressure of 45 psig. The differential can be read as about 12 psig. Therefore,

$$
\begin{aligned}
\text{cut-out} &= \text{cut-in} - \text{differential} \qquad \text{(from part (b) above)} \\
&= 45 - 12 \\
&= 33 \text{ psig}
\end{aligned}
$$

The left side of the pressure control is a high pressure cut-out designed to open on an increase in pressure to about 225 psig. The differential is set at the factory and cannot be changed in the field. It can be determined by checking the manufacturer's specification sheet for the particular control. Sometimes this information is stamped on the body of the control.

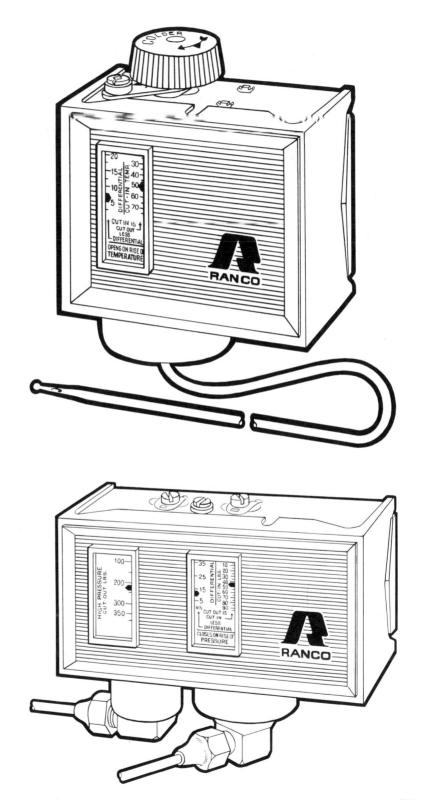

Active Cylinders	Suction Pressure (psig)	Current Draw (amps)	Capacity (%)
5	over 65	100	Full
4	63	83	80
3	61	65	60
2	59	47	40

FIGURE 4–4   Capacity control by means of cylinder unloading can be closely controlled by decreasing suction pressure. The table shows a compressor output decreases as the suction-pressure of an R-12 system unloads the cylinders of a five-cylinder compressor. Note the accompanying decrease in current draw as well.

The power company has also noticed this fact and penalizes commercial users by adding a demand charge to their normal power consumption for high spikes of power usage. By staging several small compressors, partial load conditions can be accommodated without having a single large compressor continually cycling on and off.

Beyond the power usage penalty most of the wear and tear on motors driving compressors occurs on start up. From that fact it follows that the fewer starts a compressor has to make in a 24-hour period, the longer its operating life will be. This, then, is another reason for the attempt to match the cooling capacity of a compressor with the load requirement of the building. Such capacity modulation is commonly accomplished by means of *cylinder unloading*. In this case a single large capacity compressor with a number of cylinders is used. As the cooling load in the building decreases, the refrigeration system suction pressure will also decrease. A pressure-sensitive control then begins to "unload" the cylinders. This is done by mechanically forcing the suction valve on the cylinder to remain open during the compression stroke. In this condition the pumping capacity of the cylinder is reduced to zero, thereby reducing the compressor capacity as well as the power consumption. Controls can be sequenced to unload the cylinders until only one or two cylinders are producing useful cooling. At that point, if the output of the refrigeration system is still greater than the load, the thermostat will shut down the compressor by means of a pump-down cycle. An alternative to a device that holds the suction valves open is a solenoid valve that bypasses hot gas from the discharge to the suction side of the cylinder. Usually, each cylinder to be unloaded has its own solenoid valve.

In many instances a hot gas bypass will be used either alone or in conjunction with cylinder unloading. As mentioned previously, there may be several steps of cylinder unloading. After the last cylinder is unloaded and before the thermostat in the conditioned space can shut the system down, the hot gas bypass goes to work. It is typically controlled by a regulating valve sensing the low side pressure. This valve will open as the low side pressure continues to fall after the final stage of unloading occurs. As it opens, hot gas from the compressor discharge line flows into the suction line, bypassing the condenser and evaporator and reducing system cooling capacity still further. In this way the compressor will remain running as long as possible at a very low cooling capacity, hopefully matching the heat gain of the building.

When used without cylinder unloading, the hot gas bypass valve tends to maintain a constant evaporator pressure even though the building's cooling load is falling. It does this by opening wider and wider to allow more and more refrigerant to bypass the condenser and evaporator decreasing the useful cooling effect.

4.3 WATER-COOLED EQUIPMENT

Air-cooled condensers are widely used on smaller refrigeration systems. Condensers cooled by water, however, date back to the early days of the HVAC industry and today have a very important role in many medium, and most large, installations. Water-cooled systems have generally yielded greater efficiencies when measured in Btu/watt, but the additional initial cost of installation and greater maintenance requirements have caused it to fall into some disfavor on smaller installations.

As we saw with air-cooled condensers, the control of the high side pressure at low air temperatures is mandatory. Control of the water temperature is equally important in systems using water-cooled condensers. Such systems typically call for an entering water temperature of 85°F, which should produce a condensing temperature of 105°F. With air-cooled systems the most popular high side pressure control techniques are concerned with regulating airflow through the condenser whereas with water-cooled systems the most popular systems control is the temperature of the water.

The cooling tower is a device designed to take advantage of the evaporative cooling capability of water. As the water from the condenser flows through the tower, a portion of it evaporates, cooling the remainder to some lower temperature. Just what this temperature will be depends on a number of factors such as the relative humidity, air temperature, refrigeration load, and so forth. Since more often than not the water will be at a temperature other than 85°F when it reaches the bottom of the tower, some means of control must be incorporated.

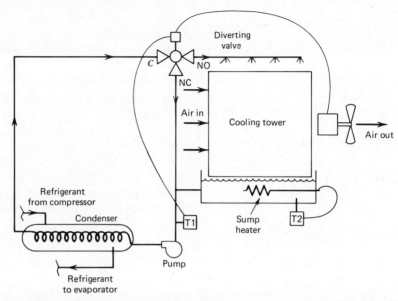

FIGURE 4–5   In this cooling tower diagram a thermostat T1 (setpoint = 85° F) will shut down the fan as the temperature falls. A continuing drop in temperature will cause the diverting valve to permit a portion of the water to bypass the tower. The tower water and bypass water will mix in varying proportions on demand of T1 to maintain 85° F. Note the sump heater that responds to T2 to prevent winter freezeup.

The control of airflow through the tower by means of dampers responding to a thermostat located in the supply line from the tower can be effective. In addition, a similarly placed thermostat can cause the tower fan motor to cycle on and off or, if both the thermostat and the motor are properly equipped, a two- or three-speed arrangement can be used. In larger towers, where multiple fans are used, individual fans are cycled in response to the water temperature.

As outdoor temperature gets quite low, down to and below freezing, a three-way valve arrangement can be used to bypass a portion of the condenser water around the tower. A mixing valve can mix return water of about 95°F temperature with an appropriate amount of water that has been cooled down by having passed through the tower.

Additional controls found on cooling towers are used for control during winter operation. During intermittent operation in below-freezing weather the water in a rooftop tower sump might freeze unless it is kept above 32°F. Electric heaters and steam or hot water coils are used for this purpose, designed to respond to a thermostat immersed in the sump.

Although cooling towers and closed circuit coolers are found on most installations of any size employing water-cooled equipment,

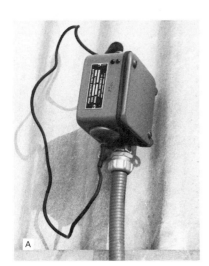

FIGURE 4–6   This cooling tower provides water to the condenser of a central chiller located in the "penthouse" equipment room at left. Low air temperatures are sensed by a controller (A), which energizes an electric heater located in the tower sump (B), to prevent tower water freezeup. Note the low water control next to the electric heater. This control will prevent the compressor from operating should the tower water fall to an unsafe level.

smaller systems may still be connected to city water supply sources. Such water varies in temperature considerably from winter to summer. Water-regulating valves are used in such systems to control the water flow rate to the condenser, much the same way as dampers control the airflow to air-cooled condensers.

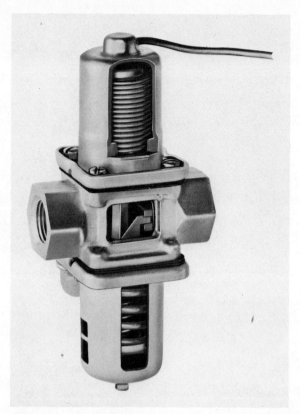

FIGURE 4–7   This water-regulating valve, shown in cutaway form, will control refrigeration system pressure under conditions of varying water temperature. (Courtesy of Controls Company of America.)

In cold winter weather the valve will throttle down the flow of water to keep the condenser pressure high. In summer, as water temperature increases, it will open, increasing the water flow rate and preventing high side pressures from increasing excessively.

4.4 CHILLER SYSTEMS

In large installations, where more than one cooling coil is required, it becomes uneconomical to use direct expansion systems. In theory DX systems can have a single condensing unit connected to several evaporators, but in practice, the trend is to use a chiller system if the distance between the evaporators is more than a few feet.

Chiller systems are refrigeration systems that cool the water to about 45°F at a central location. The water is then pumped through insulated pipes to a coil located in a duct in the same manner as hot water is pumped to a coil in a duct used for heating.

The water in such a system is called the secondary refrigerant to distinguish it from the refrigerant used in the vapor cycle. Another term for the water is *brine*. In older systems, particularly in systems where the secondary refrigerant had to be cooled below the freezing point of water, salt water was used, hence the term *brine*. Because of its corrosive characteristics it is rarely used in HVAC applications these days. Ethylene glycol or a similar anti-freeze additive is mixed with water to lower its freezing point.

The central plant of the chiller system is quite similar to the DX refrigeration system. It has a compressor, condenser, evaporator, and associated piping and controls. The condenser could be air- or water-cooled, and the problems of high side control previously described are handled in a similar manner. The main difference is in the evaporator and in the controls used to govern water temperatures.

The evaporator is called a "liquid cooler" and comes in a number of shapes dependent upon its application. In HVAC systems it is quite often cylindrical in shape with tubes in the cylinder carrying the primary refrigerant and the water surrounding the tubes. Such an arrangement uses a thermostatic expansion valve to control the primary refrigerant flow. An alternative system, called a "flooded system" has the water flowing through the tubes and the refrigerant surrounding them. Such a system uses a low side float to control the primary refrigerant flow and uses hundreds, even thousands, of pounds of refrigerant in its charge.

Chiller systems are commonly found in sizes in the hundreds of tons of refrigeration. It would be advantageous to use a control technique other than the two-position on-off method, since the compressor motors are so large. Much of the complexity of the chiller system is a result of having a control system with the capability of matching the output of the equipment with the cooling requirement of the building.

At design conditions the typical chiller receives water from the building at 55°F and should cool it down to 45°F. The terminals throughout the building are in turn designed to receive chilled water at 45°F and, in the cooling process, the water is then warmed to 55°F. The most basic chiller control is a thermostat in the leaving side of the chiller that shuts down the compressor when the water temperature drops to 45°F. We have discussed before why this on-off action is undesirable.

To extend the "on" time as long as possible, unloading is built into most chillers. Unloading the chiller compressor is accomplished in ways quite similar to those previously discussed. The chilled water thermostat can be designed with several stages to unload the cylinders by means of bypass solenoid valves. The compressor may also be unloaded in response to dropping suction pressure. Larger chillers use centrifugal and screw-type compressors. The control of such compressors is rather simple and

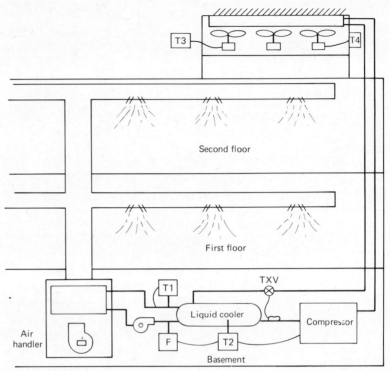

FIGURE 4–8   In this chiller installation an air-cooled condenser is used. Head pressure control is maintained by fans cycling in response to T3 and T4 outdoor air thermostats. The compressor is energized when supply water temperature rises if the water flow switch is closed and if the low temperature safety, T2, is closed.

more closely approaches proportional control. Whereas the reciprocating compressors used on smaller chillers are unloaded in increments, that is, by cylinders, the centrifugal and screw compressors usually found on the larger systems are unloaded by controlling the amount of refrigerant entering the suction side of the compressor. A damper positioned in response to the chiller water temperature or refrigeration suction pressure is commonly used on the suction side of the compressor. A slide valve allowing more or less refrigerant to enter the screw compressor is also positioned by chilled water temperature or suction pressure.

Although the simple chiller system calls for a chilled water supply temperature of 45°F, increasing concern with energy conservation has resulted in the reevaluation of this requirement. A building with a relatively light load really does not need 45°F water to cool it. Water at 48°F or perhaps a bit higher could still do an adequate job of cooling, but it might take longer to do the job. A

FIGURE 4–9   This chiller control panel combines pneumatic and electrical controls to maintain water temperature at desired levels and to monitor the operation of the equipment so that in the event of an unsafe condition the equipment will shut down.

relatively new scheme provides for the set point of the chilled water thermostat to be reset by an outdoor controller. As the outdoor temperature falls below the design temperature, the set point of the chilled water thermostat increases. Although the same cooling job is being done whether the water temperature is at 45 or 48°F, with a higher water temperature the compressor is operating more efficiently at a higher suction pressure and the piping losses are also somewhat reduced.

Beyond controlling the water temperature, there is a safety requirement that must be met, namely, to insure that the water in the liquid cooler never freezes. The expanding ice will rupture the evaporator, a calamitous occurrence. Two controls are commonly used, both having the same purpose, shutting down the compressor should conditions that might lead to a freezeup occur. A thermostat with its sensing element immersed in the water within the liquid cooler can sense if the temperature falls too low, a common cutout point being about 35°F. Such a condition might occur if the water flow through the liquid chiller dropped too far below the design point. A flow switch installed in the chilled water line will cut off the compressor if the water flow rate falls to an unsafe level. It is usually wise to have at least one of these controls be of the manual reset type.

FIGURE 4–10   This flow switch will open an electrical circuit in the event that water flow through the line is less than required for the safe operation of the equipment. Note the arrow indicating the proper flow direction for proper installation. (Courtesy of ITT Fluid Handling Division.)

4.5 HEAT PUMPS

A literal interpretation of the term *heat pump* would lead one to believe that it is a device designed to move heat from one place to another. This broad definition is used to describe a number of systems that do exactly that, each of them having some very major differences in how they are put together. At this point we consider the heat pump that employs a reverse cycle refrigeration system; a bit later we shall discuss some larger systems using other techniques of heat pumping.

To a very large number of people in the HVAC industry the term *heat pump* has only one meaning, that is, it is a system that provides cooling by means of a direct expansion refrigeration system and is also capable of providing heat by reversing the flow of the refrigerant. In effect, the heat exchanger that acts as the evaporator during the cooling cycle becomes the condenser during the heating cycle. Most heat pumps of this type are used in residential or smaller commercial applications and are of the air-to-air type. This means that they reject heat to the outdoor air during the cooling cycle and absorb heat from the outdoor air during the heating cycle. The indoor coil is installed in a duct through which air, requiring heating or cooling, flows on the way to the conditioned space.

The heat pump differs from the straight cooling unit in several ways. A means of reversing the flow of refrigerant must be provided. This is accomplished by the four-way valve, or reversing valve, which responds to a signal from the thermostat in the conditioned space. The flow through the compressor is always in

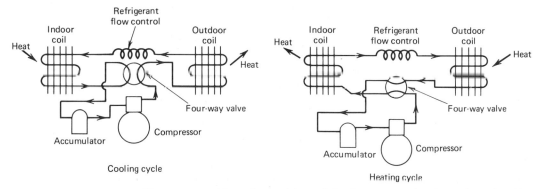

FIGURE 4-11   Heat pumps. Note the position of the four-way valve in each cycle. The accumulator has been added in recent years to help the compressor withstand the liquid sludging that can occur on startup in cold weather. Even in very cold weather the outside air has heat that can be extracted and moved indoors.

the same direction; however, after leaving the compressor, the four-way valve directs the refrigerant flow either to the indoor coil for heating or to the outdoor coil during the cooling cycle.

As the outdoor temperature drops during the heating cycle, two things happen. First, the refrigerant temperature drops, leading to frosting of the outdoor coil, which, if allowed to continue unchecked, will cause the system's heating ability to drop to practically zero. Second, the output in Btu/hr also drops. This requires that on cold days supplemental heat, in the form of electric heaters, be turned on to provide the difference in heat between that required and that produced by the heat pump.

To prevent excess frost accumulation, a defrosting cycle is provided. The defrost control senses a buildup of frost either by sensing increased resistance to airflow, that is, a greater pressure drop through the outdoor coil, or by sensing the difference between the outdoor air temperature and refrigerant temperature. This difference will increase as the outdoor coil becomes increasingly blocked by frost. One type of defrost controller merely initiates a defrost cycle at fixed time intervals whether the coil is blocked or not. Usually, the method of defrost is to put the system back into the cooling mode and let the hot refrigerant vapor pass through the outdoor coil for a few minutes. To counteract the cooling effect felt indoors, the electric heaters are used to reheat the air. Termination of the defrost cycle is usually accomplished by a thermostat sensing the coil temperature and terminating the cycle when the coil temperature rises to a frost-free level. A time limit is commonly used as a backup in the event that the thermostat fails or if the weather is so cold that the defrosting temperature just never increases to a high enough level.

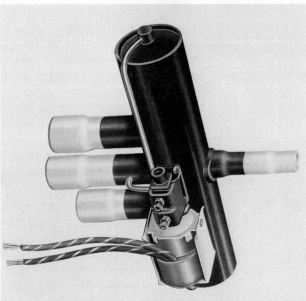

FIGURE 4–12   Two types of four-way reversing valves found in heat pump units. These pilot-operated valves employ a solenoid operated pilot valve to direct pressure to the slide contained in the body of the main valve. The action of the solenoid valve is controlled by a thermostat, either remote-mounted or built into the equipment. (Courtesy of Ranco Inc.)

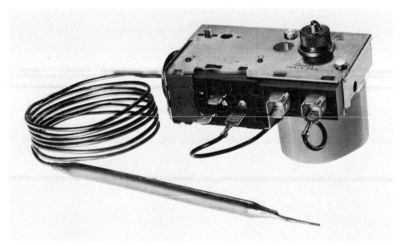

FIGURE 4–13   The sensing bulb of this heat pump defrosting control is mounted on the outdoor coil. A timer motor periodically puts the heat pump into defrost as long as the outdoor coil is below a prescribed temperature, typically 26° F. As the coil heats up, causing the ice to melt, and rises above 55° F, the defrost cycle is terminated. If the temperature never exceeds 55° F, the cycle will be terminated by the control in about 10 minutes. (Courtesy of Ranco Inc.)

The supplemental heaters are usually controlled in response to the outdoor temperature. In some units, as the outdoor temperature drops below a level determined by the manufacturer, the refrigeration cycle is shut down and the heaters take over. At that point any call for heat from the thermostat in the conditioned space results in the electric heaters being energized rather than the four-way valve being reversed and/or the compressor coming on. In some systems two stages of electric heaters are used. When the outdoor temperature drops to say 30°F, the first stage only may be energized by the room thermostat. An additional decrease to 20°F may bring the second stage into operation. This is an energy conservation scheme that prevents all electric heaters from being used when the outside temperature clearly indicates that this is not necessary. Such a situation might arise when the system is started in the morning and a rapid increase in temperature is desired.

If this sounds complicated compared to straight cooling units, it is. Heat pumps have been carefully designed both electrically and in refrigeration circuitry to operate for periods as much as five times longer during the year than cooling-only systems. They are being widely used, particularly in mild climates, with increasing reliability.

The more popular versions of air-to-water and water-to-water heat pumps are usually found as part of a large system, where energy conservation is a major system goal. We shall look at such applications when we deal with energy conservation techniques.

DISCUSSION TOPICS

1. How does a cooling anticipator work?
2. Describe some techniques of high side pressure control in a refrigeration system.
3. How does a pump-down cycle work?
4. Why is cylinder unloading used?
5. Describe the pro's and con's of water-cooled and air-cooled air conditioning equipment.
6. What is a secondary refrigerant? Under what conditions is it used?
7. What is the purpose of a flow switch in a chiller system?
8. Describe the difference between an air-to-air and a water-to-water heat pump.
9. Why is defrosting required on a heat pump? How is the defrost cycle initiated?

Chapter Five

Distribution of Cooling

Although the expression "distribution of cooling" is not technically accurate, since the process of cooling is actually the removal of heat from a conditioned space, we shall use it anyway. In describing how a conditioned space is cooled, we deal with the problems of sending cool air into the space. This air is cooled by either passing it over a DX coil or a chilled water coil at a central location and then distributing it throughout the building by ducts or it is cooled at terminals located in or near the conditioned space and discharged directly into the space. In this chapter we consider some of the techniques used to accomplish this.

5.1 DIRECT EXPANSION SYSTEMS

In a small residential or commercial application there is no effort made to control the temperature of the air flowing through the ducts to the cooling terminal. This temperature is related to the return air temperature, outdoor temperature, or building cooling load, and is uncontrolled. In larger systems, however, there is an effort made to control this temperature to match the building load and therefore to decrease the number of on-off cycles of the equipment, thus preventing unduly cool drafts in the conditioned space.

In heating applications, control of the coil temperature was accomplished by varying the water flow or water temperature. In cooling applications, the coil temperature is usually varied only slightly. The reason for this is the requirement for dehumidification that usually accompanies the requirement for cooling. Until rather recently, the coil temperature was not controlled at all. However, increasing concern about the use of energy has led to newer systems that provide a measure of coil temperature control within closely defined limits, these limits being defined by the dew point temperature range of the particular locale.

Since coil temperature is not too commonly used, the only feasible way of controlling the air temperature in DX systems is to pass a portion of the air over the coil and to bypass the coil with the balance of the air. This is the face and bypass damper arrangement discussed in the section on heating. A thermostat in the cooling duct or in the conditioned space will position the dampers to provide the proper ratio of conditioned and unconditioned air to meet the cooling requirements. Other control configurations previously described can be used, for example, the master–submaster system with an outdoor thermostat or a room thermostat controlling the set point of the discharge air thermostat.

Note that since we are not controlling the coil temperature, the air passing over the cooling coil will be dehumidified. As the cooling load in the conditioned space decreases, more air will pass through the bypass and less through the coil. The DX system must have an unloading capability built into it. Such unloading systems were described in Chapter 4. In that description unloading was required due to cooler and cooler air passing over the DX coil. In the case of face and bypass control the air is still relatively warm as it passes over the coil, but there is less of it. The result is the same: a lower suction pressure and a colder coil, which will freeze up in the absence of unloading.

Although coil temperature has not been controlled much until recently, some measure of control can be achieved in DX systems and it is probably worth a word or two at this point. A hot gas bypass valve, sensitive to suction pressure, is used. As the suction pressure falls below the set point of the valve, it opens allowing hot gas from the discharge line to enter the suction line maintaining low side pressure. As the space thermostat indicates that less cooling is required due to a dropping space temperature the face dampers close and the bypass dampers open. The resulting falling suction pressure then brings the hot gas bypass valve into play. In compressor systems using unloaders, the last stage of capacity control is accomplished with the hot gas bypass valve.

5.2 CHILLED WATER SYSTEMS

More often than not, control of the cooling effect in ducted systems employing chilled water duct coils is obtained by a technique other than water temperature control. As mentioned earlier, one desired effect of the cooling process is dehumidification. Coil temperatures above the dew point are liable to result in cool, damp buildings, a most unsatisfactory situation. Recent energy conservation innovations have produced control systems that adjust the water temperature at the chiller in response to the outdoor temperature and humidity level. On dry days less dehumidification is required so that the water temperatures can be permitted to increase a few degrees.

Face and bypass dampers can be provided to give some measure of dehumidification with minimum cooling. A cold coil is maintained, but decreasing amounts of air pass over the coil as the

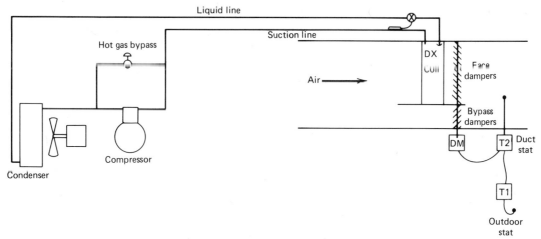

FIGURE 5–1   Control of the coil temperature can be achieved as the airflow through the coil decreases through a combination of cylinder unloading and hot gas bypass. Then T1 and T2 act to control the duct temperature while the hot gas bypass and cylinder unloaders work to lengthen the "on-time" and minimize short cycling of the equipment.

conditioned space temperature drops. If humidity conditions become excessive in the conditioned space even though the temperature is dropping, more air can be directed through the duct coil to remove moisture and a reheat coil can be used downstream of the coil to bring the air temperature back to a comfortable level. Increasingly, energy conservation procedures are suggesting that reheat not be employed unless very close tolerances of humidity level are required. Typical examples of this might be a computer room or perhaps a space in which an industrial process requiring strict humidity control is being carried out.

Since the water temperature is usually kept relatively constant, the remaining variables that can be controlled are the water flow rate and air volume. Proportional flow of chilled water can be used to control the supply-duct air temperature. A problem with this approach is that as the water flow rate decreases, the spread between the entering and leaving water temperatures to the duct coil increases. Typically, a coil is designed to provide a rated cooling capacity with entering water at 45°F and leaving water at 55°F. By throttling down the water flow, the temperature of the water leaving the coil will increase. Less coil surface will be available for dehumidifying as the exit water temperature increases. In parts of the country where dehumidification is not a major concern, such a system can be used effectively.

Increasingly variable air volume systems are being applied. The airflow in the duct will vary in response to a thermostat in the conditioned space. This lends itself to combination heating/cooling systems and we shall discuss such VAV systems in detail in Chapter 7.

The extreme in variable volume control, that is on-off control, is seen in small fan coil systems. The room thermostat controls the action of a fan blowing air across a coil. On a call for cooling the fan runs; when the thermostat is satisfied, the fan shuts off.

5.3 DUCT SYSTEMS

Most duct systems in use today are low velocity systems, where air velocities of less than 2000 feet per minute are typical. High velocity systems had achieved some popularity in applications where the space for ductwork was minimal. Since high velocity systems could do the job with smaller ducts, there seemed to be an economy that could be realized. In practice it turned out that such systems velocities often substantially greater than 2000 feet per minute, were noisier, required stronger ductwork to contain higher pressures, suffered from greater duct losses through air leakage due to these higher pressures, and cost more to operate because of higher fan horsepower requirements. Today, unless space limitation is overriding concern, low velocity systems predominate.

Material of duct construction is mainly galvanized sheetmetal. In locations where corrosion might be a problem, aluminum ductwork will often be found. Pressed board made of fiberglass has also been used in a growing number of applications. A major benefit of such board is that no additional insulation need be wrapped around the duct, unlike metal ducts that must be wrapped with insulation to minimize heat loss or gain, and perhaps more important in the case of cooling ducts, to prevent sweating of the duct as it passes through areas of uncontrolled temperature and humidity. A drawback of fiberglass ductwork is that it requires special care and reenforcing when installed in areas where it is likely to be stepped on or subjected to high pressure loading. Typically, metal ducts do not have that problem.

Most emphasis in designing duct systems is given to the supply ductwork. This is the network of main and branch ducts carrying the conditioned air from the main equipment room to the conditioned space. Air velocities are low enough to provide a quiet system, yet high enough to deliver air to the terminal with sufficient pressure so that the air is introduced to the room evenly and with enough velocity to obtain good air distribution within the space.

Often neglected is the return air system that must remove the air from the conditioned space and return it to the equipment room for conditioning and recirculation. A major problem, often attributed to improper control, is inadequate return air flow, particularly from the extremities of the air distribution system. In some cases a return air fan is installed in the system to overcome the resistance of the return ductwork. This fan is intended to work with the supply fan and the controls for both must be designed to accomplish this goal.

A good way to test the adequacy of the return system is to measure the air pressure at the intake of the main supply fan. If it is a large negative pressure, the return is being restricted. In some systems just attempting to remove an access panel at the intake side of the fan will give a good indication of the adequacy of the return. If the panel can only be removed with difficulty and if it is strongly sucked back into position when released, then the return air is being restricted. This is as true on small residential systems as it is in large commercial structures. Often the solution is nothing more than adding larger returns, but it might be as involved as installing a return air fan.

5.4 PIPING   Pipes are used to carry chilled water or hot water or steam from the central plant to the duct coils or terminals. In smaller systems copper tubing up to about 4 inches in diameter is used. Larger systems use steel pipe with threaded or flanged fittings. A major requirement is that the chilled water, leaving the chiller at a temperature of about 45°F, reach the terminal unit or coil with little or no increase in temperature. Although theoretically it could be argued that any heat absorbed by the water as it moves through the building is useful cooling, in actual fact this may not be the case. In any event, control of the cooling process occurs at the coil or terminal and any heat picked up prior to that is considered a waste of energy. For this reason pipes carrying chilled water, and by similar reasoning pipes carrying hot water or steam, are insulated. Another reason for insulating chilled water pipes is to prevent moisture in the air from condensing on the pipe surface, which is usually at a temperature below the dew point. Chilled water pipes should not only have enough insulation to minimize heat pickup or loss but also to form a vapor barrier covering the insulation to minimize this condensation.

Instrumentation of the piping system as well as the duct system is particularly useful. Instruments incidentally can be considered devices that measure control variables and present this measurement in some sort of display, like a thermostat scale or a pressure gage. Controls not only measure the variables but take action to cause changes in the variables. Note that controls and instruments are *not* the same thing. Pressure gages before and after the pumps and duct coils enable the controls person to determine at a glance whether or not the system is operating properly. Of course this assumes that the controls person is able to recognize proper and improper readings on the instruments. This is another good reason why the controls person should be the most widely experienced

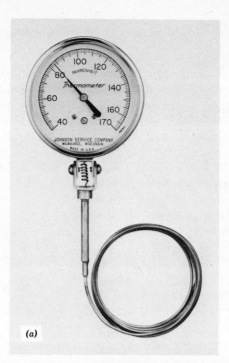

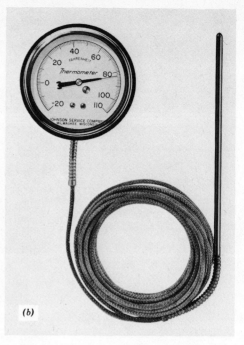

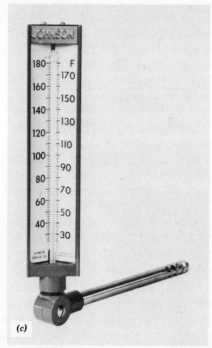

FIGURE 5–2   Three thermometers that are used for measuring air temperatures in ducts. Instrument (a) has an 8-foot averaging bulb and must be mounted very close to the point of measurement. Instrument (b) may be mounted remotely because only its bulb need be inserted in the airstream. Instrument (c) is a vertical scale immersion type of thermometer that must be mounted on the duct wall. (Courtesy of Johnson Controls Inc.)

FIGURE 5–3   Several pressure gages and thermometers installed in an equipment room. In addition to these instruments, several remote sensing instruments that feed information back to a central control room may be included.

and most capable member of the building team. Manufacturers of pumps and coils provide pressure information as a guide to performance. A pump is designed to operate with a specific pressure increase from inlet to outlet at a particular flow rate. The pressure decrease as water flows through a coil is specified by the coil manufacturer for a particular flow. This information is usually available in the form of graphs of pressure drop versus flow rate. If the graphs are available and the gages are already installed, the building operator, service person or controls startup person can tell at a glance if the system is operating the way it was designed.

The piping should also have thermometers installed at similar locations so that the system's performance can be analyzed. Economy-minded designers leave out such instruments, and miserly building operators may not replace those that go bad. Although the information can still be obtained by the person on the job by personally installing the appropriate instrumentation, the chances are that it will not be done. An important maintenance procedure is to record data on a daily basis so that trends can be observed. In the absence of instrumentation the gradual system changes that eventually lead to a major failure cannot be observed. Thus a "saving" has been turned into a loss. If computerized supervisory systems have only one thing to offer, and detractors of such systems say that is so, it is that they can automatically monitor pressures and temperatures in piping and duct systems throughout a vast building complex and alert operating personnel when system performance starts to slip, and *before* major problems arise.

One last word to the controls person about piping. Incredible as it may sound, mistakes can be made at any point in the construction process from the design stage through installation and startup. It is possible for a beautiful system, complete and ready to go, to have pipes connected improperly. If the line is ¾-inch copper tubing, it can readily be corrected; however, if it is a 12-inch steel pipe leading to the inlet of a cooling tower that should have been connected to the sump, it cannot be so readily corrected. But it must be done. *Not all control problems are problems with controls.* The controls person must have a sound background in *all* aspects of system design, not just controls, so that problems in any area can be spotted and solved.

5.5 TERMINALS

The terminal is the device that introduces the supply air into the conditioned space. A number of techniques are available for what would seem to be a simple task. Square and circular ceiling diffusers are intended to introduce the air into the space so that its velocity parallels the ceiling. The air moves along the ceiling to the walls and hopefully drops in a draft-free manner to minimize any discomfort. These diffusers usually have dampers provided to aid in adjusting the airflow. Although a measure of control can be obtained by using these dampers, the main control of the airflow should be accomplished with volume dampers in the ductwork itself. As the diffuser dampers are moved toward a closed position, air velocity noises can become quite annoying. Most noises generated by the volume dampers located in the duct are muffled by the ductwork itself. Therefore, the proper adjustment of these dampers is quite important, particularly if a large area is being served by one thermostat. Because the space thermostat only senses the temperature where it is located, for this to be an accurate sample of what is happening throughout the space, even air distribution is required. In some cases hot or cold spots may exist in the space so that diffusers serving these areas may have to be adjusted to provide more or less airflow than others in the system.

Sidewall registers are located to direct streams of supply air into the conditioned space. Air must be directed so that it does not cause uncomfortable drafts. Perhaps more important, the air must be directed so that it does not hit a thermostat. The supply air is usually colder (or in the case of heating, warmer) than the conditioned space. A stream of supply air directly hitting the thermostat will invariably cause erratic operation of the system.

Slotted diffusers are used in decorative lighting or special ceiling effects. Long, narrow bands of conditioned air flow into the room.

If they have been selected properly, an adequate amount of air will enter the room. Uneven temperature conditions may exist if the air velocity across the entire length of the slotted diffuser is not uniform. Usually, damper adjustments can be made to provide such even distribution.

Slotted or longitudinal diffusers are sometimes used in the floor adjacent to glass doors or floor length windows and along a sill under standard windows. Again, air distribution is important and every effort should be made to keep the diffusers unobstructed. Window sills are notorious for collecting books and papers, flower pots, and so forth, as well as having curtains hanging close to them. Such obstructions will defeat the effect of the terminal and cause the occupants of the building to blame the control system.

In variable volume systems the terminals may be equipped with dampers that move in response to the space temperature. Balancing the maximum airflow of the system is accomplished by manual volume dampers in the ductwork, whereas temperature control in the space is accomplished by motorized dampers responding to space thermostats.

An induction unit is a terminal very similar in design to a fan coil except without the fan. A central air conditioning unit sends conditioned air to the terminal much the same as to any diffuser. The air passes through a nozzle arrangement within the induction unit. The velocity of this air, called primary air, passes through the nozzle and creates a negative pressure drawing in room air (secondary air) at the lower front of the unit. Primary conditioned air and secondary room air mix and then pass through a coil where reheating or cooling may take place. Motorized dampers are sometimes used to regulate airflow in response to room temperature. In some designs two-way or three-way valves are also used for coil temperature control. The induction system is really a combination of ducted and piped air conditioning systems.

5.6 INCREMENTAL COOLING

A major benefit of having a central cooling plant is the concentration of the major components at one location. Supporters of incremental systems claim that a major shortcoming of the central system is that the system's response to partial cooling loads is not precise and also that the failure of one or two large compressors, or other key components, affects the entire building. Incremental systems are characterized by each area of the building having its own cooling unit. The term *packaged terminal air conditioner* (PTAC) has been used to describe such incremental units. Such units may be nothing more than a room air conditioner or may come in a decorative cabinet and include heating means using electricity, hot water, or steam, controlled by a wall-mounted thermostat.

The idea of having each conditioned space served by its own total air conditioning system has gained support in applications such as hotels, motels, apartment buildings, and office buildings where the individual spaces are next to an exterior wall so that through-the-wall air-cooled units can be used.

Another incremental system uses a large number of small capacity water-cooled heat pumps. This scheme is neatly applied to buildings where a number of interior areas must be served as well as perimeter offices. The heat pumps in the interior operate in the cooling mode, absorb heat from the core of the building, and then transfer it to the cooling water. This warmed water flows to the perimeter units that might need heat. These perimeter units operating in a reversed cycle, or heating mode, absorb the heat from the water and reject it to the conditioned space. In this system heat is being transferred from where it is not wanted to where it is wanted. When properly maintained, this is a particularly good energy-conservative approach to conditioning buildings in the spring and fall, when heating and cooling may be needed simultaneously in a large building.

A similar concept uses water-cooled incremental cooling units with hot water heat. In mild weather the heat rejected to water-cooled condensers can be pumped to the cool parts of the building and passed through the hot water heating coil of another unit. Air passing over the coil removes the heat from the water and carries it into the room. Such a system does not provide heating and cooling all year round but does enable the cooling and heating seasons to be extended through the "in-between" days of fall and spring. The compressors used in these units only work part of the year rather than year round as heat pump compressors must.

DISCUSSION TOPICS

1. How can control of coil temperature in DX systems be achieved?
2. How can air temperature in a chiller system be controlled?
3. What is a low velocity duct system? Why is this system popular?
4. What is the major drawback of a fiberglass duct?
5. What is one indication of an inadequate return system in a ducted HVAC application?
6. What is the difference between an instrument and a control?
7. How might instruments installed in an HVAC system be valuable?
8. Why are incremental systems popular in some applications?

Chapter Six Building Requirements

The air conditioning of a building requires control of the temperature, humidity, and air cleanliness and distribution. One of the interesting parts of controlling the air conditioning process is that the requirements for air conditioning are not uniform throughout the structure. An individual walking through a building with no air conditioning system at all on a cold winter day would find that in one part of the building he or she might be very cold, in another moderately cold and in still another uncomfortably warm. This occurs because the amount of cooling or heating (the load) throughout a building is not uniform. Any effort to assume that the load is uniform and to design an air conditioning system based on this premise will result in a less than perfect design.

Increasingly, the building is considered to be composed of two areas: (1) the perimeter, that is, any area within about 20 feet of an exterior wall; and (2) the core, that area more than 20 feet from an exterior wall. In this chapter we shall examine the requirements of each of these areas as well as some common techniques for meeting these requirements.

6.1 THE PERIMETER

The amount of heat that leaves a building in winter or enters the building in summer is the external load of the building. The perimeter system is intended to handle this external load as well as that part of the internal load generated within about 20 feet of the exterior walls. The main characteristic of the perimeter load is that from time to time it changes. The change may be a result of the sun's traveling across the sky, winds shifting or varying in intensity, cloud formations covering the sun periodically, or just changing outdoor temperature. In any case, whether in the cooling or heating season, the load on the building perimeter varies.

Ideally, perimeter air conditioning systems have a capability to cool *and* heat, a capability that is available, if not 365 days a year, at least over as long a period of the year as the requirement for both heating and cooling is likely to occur during a day.

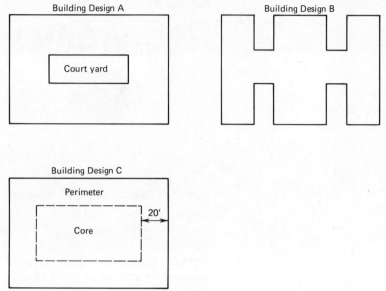

FIGURE 6–1   Building designs A and B are plan views that show how early office buildings were built with large exterior wall areas. Design C shows the boxlike design used today that requires mechanical ventilation for core areas.

Perhaps the simplest perimeter system employs fin tube radiation using steam or hot water to provide heating only. Fan coil units with heating or cooling or both are also applied here, as is the induction system. Increasingly popular is the PTAC, in which an air- or water-cooled refrigeration package is combined with a steam, hot water, or electric heating coil to provide individual control around the perimeter. Heat pump units have also found a place in this application.

6.2 THE CORE   Since by definition the core of the building is that part 20 or more feet from an exterior wall, we can characterize the core load as being stable. The major load in the core is a cooling load and consists of interior sources such as people, lighting, and equipment, the greatest of these usually being heat generated by lights. There should be no requirement for heating in a core system, but since outside ventilation air is always being brought in, at least while the building is occupied, some sort of heating is usually provided.

Core systems in nonresidential buildings are almost invariably ducted. In residential buildings such as apartment houses and hotels, a combination system may be used with a ducted system serving hallways and common areas. Outside air is brought into

FIGURE 6–2   This photograph, taken during the late construction stages of an office building, shows the core and perimeter systems used. The duct will eventually be tapped with branches to supply conditioned air to offices in the interior; the fan coil units shown against the wall will handle perimeter offices.

these areas, and infiltration is relied on to carry the fresh air into the individual living areas. Exhaust fans in these areas may help the ventilation process.

6.3 ZONING   Although the building may be considered to be composed of a core and a perimeter, this is generally not a sufficient breakdown for designing the HVAC system. At any one time the load on different points of the perimeter may be different and, depending on usage, the load at different points in the core may be different as well. It is useful to attempt to divide areas in the building into blocks having fairly similar loads at a given time. These blocks are called *zones*.

The least expensive way of conditioning a building, at least from the viewpoint of initial cost, is to assume that the entire structure is one zone. This is commonly done in single family residences and small commercial buildings. A single thermostat at what it is hoped is a location sensing the average building temperature controls the heating/cooling equipment in the building.

The result is that those occupants near the thermostat are comfortable while those remote from the thermostat may or may not be comfortable depending upon how close the local temperature is to the temperature at the thermostat location. Generally, we consider a zone to have a single thermostat. The larger the zone, the less likelihood that all occupants will be comfortable; the smaller the zone, the greater the likelihood. The smaller the zone and the more zones in a building, the more complicated the control system and therefore the more expensive the system.

Once the decision is made that a single zone approach will not be used, and this decision is very closely tied to the building function, it becomes important to accurately break the building down into appropriate zones. Some considerations in this breakdown are expense, solar effect, prevailing winds, the number and characteristics of the occupants, the activities of the occupants, machinery and lighting, the perimeter or core location, and open area or small cubicles in the zone.

The intent is not necessarily to break the building down into blocks having the same cooling or heating load but rather to identify the zones that are uniform in load requirements and in which the occupants can be uniformly satisfied by air introduced at a particular temperature and flow rate. On paper this is a lot easier than in practice, and in many installations the control system is held up to ridicule when, in fact, the selection of zones and the control location is actually at fault. Once a system is built, the changes are costly. The cheapest changes are those made with an eraser during the design stage.

Figure 6–3 gives some examples of zone selection. Usually, wide-open areas can have fewer zones than areas containing individual cubicles. Combining perimeter and core areas into a single zone often leads to problems. Combining areas with active people and areas with sedentary people in a single zone will also produce complaints. Perfect zoning is the goal, but obviously constraints of budget as well as differences in the attitudes and responses of the occupants will yield a less than perfect result.

Equipment and systems have been developed to handle the simplest single zone as well as the most complex multizone application. Before looking in detail at complete systems, however, we shall take a closer look at some of the smaller subsystems that are found in larger HVAC installations. The expression "You can eat an elephant if you cut it into small enough pieces" is particularly true with respect to large complex HVAC systems. The ability to identify smaller subsystems within the larger, complex system will be a great aid in understanding the workings of such installations.

6.4 SUBSYSTEMS

Any large central air conditioning system has several or all of the following capabilities built into it: ventilation, an economy cycle, air cleaning, preheating, heating, cooling, reheating, dehumidification, humidification, static pressure control, and fire safety.

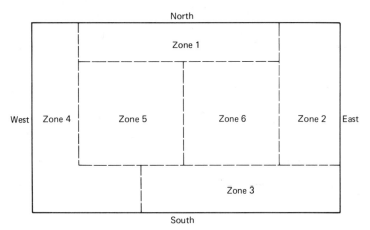

FIGURE 6-3   The floor plan of this building shows a zoning scheme based on exterior wall exposure, core location, usage, occupancy, and so forth. The zones need not be identical in size but rather should have uniform loading within the zone. Ideally, one thermostat in the zone will be able to maintain a uniform level of comfort throughout the zone.

6.4.1 Ventilation   Building codes specify the amount of outdoor air that must be brought into a building to provide for human comfort. Generally, a minimum of 5 cubic feet per minute (cfm) per person is recommended, but frequently a higher figure is specified. Increasing energy consciousness is reducing this figure toward the minimum. In industrial or some hospital applications significantly more than the minimum is required. In some extreme cases 100% outside air is specified. In such an application all the air passing through the air conditioning system is exhausted from the building and none is recirculated.

In smaller structures the ventilation requirement is met by infiltration, that is, the natural passage of air through the minute openings in the structure. In larger buildings the air handling system must supply this air.

Figure 6-4 shows one duct arrangement used for mixing outside air and return air. When the supply air fan is energized, a signal is sent to the outdoor air damper operator causing it to open the damper. How far it opens is fixed by the linkage connecting the operator to the damper, or by the number of degrees travel of the motor (in the case of electrical or electronic systems) or the length of stroke (in pneumatic systems). An important bit of information for any controls person to learn is how far the damper should be open. If it is open too far, energy will be needlessly used to condition the air; if not far enough, the air in the building may become stuffy and unpleasant. Although building codes usually specify the amount of outside air required for a building, an

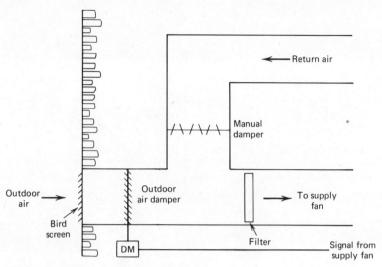

FIGURE 6–4   Minimum outside air duct arrangement.

FIGURE 6–5   Outdoor air dampers can be either parallel blade (above) or opposed blade type (right) depending upon the degree of accuracy of control required in the design. In a two-position application the parallel blade is favored over the opposed blade; in modulating applications, such as the economizer cycle, the opposed blade is more commonly used. (Courtesy of Honeywell Inc.)

interesting approach is finding its way into the HVAC market: an electronic device analyzes the air in the return duct and determines the level of carbon dioxide in the air. If it is excessive, the device sends a signal to open the outdoor air damper more, allowing increased amounts of fresh air into the building. Presumably, the cost of this device will be paid for by the energy saving associated with having only the proper amount of fresh air entering the building at any one time, thereby reducing waste.

Notice the manual dampers in the return duct. These are set to restrict the flow of return air so that the pressure in the duct may be low enough to draw in outside air. If such dampers are not present or if they are improperly set, the volume of outside air will be uncontrolled no matter what the setting of the outdoor air dampers. It is the function of the testing and balancing people, employed after the system has been completely installed, to set the dampers to allow for proper airflow.

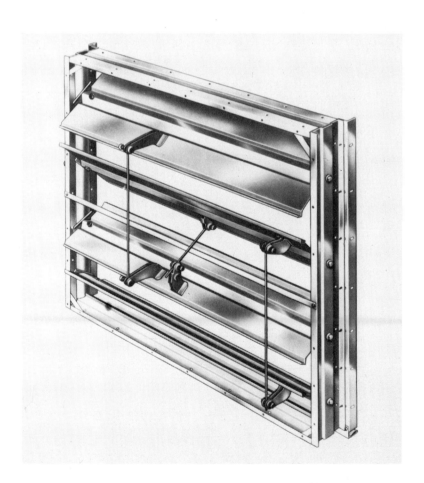

In Figure 6–6 a relief fan has been added. If the requirement for outside air is large, the pressure in the building may become excessive. This can be likened to blowing air into a paper bag. The pressure in the bag increases and eventually the amount of air that can be blown into the bag drops to zero. So too with a building. Some air will "exfiltrate" or leak out through normal building openings. However, if air is pumped in faster than it leaks out, the pressure will rise and the ventilation airflow will drop. In addition to insufficient air, such pressures make doors difficult to open and will also cause excessive noise as the air leaks out around windows and doors at a high velocity. The relief fan permits control of the air. Usually, it is designed to exhaust a bit less, say 10% less, than the amount of fresh air brought in. A slightly positive building pressure is desirable because it keeps street dust and dirt from entering through the doors and window openings.

An alternative to a relief fan is a set of dampers located in a position where they can permit any built-up pressure to be relieved. The choice of a relief fan or relief dampers depends upon the building design, air volume, openness of construction, and other factors the design engineer must consider. Note that control of the relief mechanism is tied to the supply fan. When the fan is on, the outdoor air dampers and the relief dampers will open.

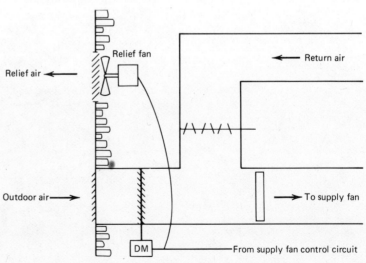

FIGURE 6–6   A relief fan energized along with the outdoor air damper to keep a slight positive pressure in the building. The quantity of ventilation air and relief air is fixed.

FIGURE 6–7   The set of louvers at the left of this mechanical equipment room permit fresh air to be drawn into the building. Those at the right permit stale building air to be exhausted. On the opposite side of the louvers is a birdscreen and a set of dampers that are automatically controlled to permit air to enter or leave.

6.4.2 The Economy Cycle   Before the age of mechanical air conditioning systems with automatic controls, the main technique of comfort control was to open the window and let in the cool outdoor breezes. The use of air conditioning is now widespread; in fact, the windows in many buildings today are designed so that they cannot be opened. Since the mechanical equipment has a heating and cooling capability, there is no reason to open windows and allow dust, dirt, and unpleasant street noises to enter the building. At least that was the thinking that led to many of the building systems in existence today.

Depending upon the orientation of a building, its window area, its location, its internal loading, and other factors, it is quite conceivable that a cooling load exists in a core area and also around parts of the perimeter on those days when the air temperature is as low as 35 or 40°F. In the days of cheap power it may not have been a cause of concern to have the refrigeration equipment in operation on such a day. But today, in the era of the "energy crisis" it most certainly does cause concern. The "economizer" or "economy cycle" is one solution to the problem of energy conservation that has become quite popular.

Figure 6–8 shows the same ventilation air ductwork previously described. However, a set of dampers has been added to the relief fan and the return air dampers have been motorized. The controls that have been added are of the modulating type. Where previously the dampers opened to a fixed position when the supply fan was energized, now the dampers will modulate to maintain the air temperature entering the air handler.

The proportional thermostat T1 is designed to maintain an air temperature between 50 and 55°F going to the air handler. This is considered cold enough to provide a desired cooling effect, yet warm enough not to cause uncomfortable drafts. Thermostat T1 does this by mixing appropriate quantities of return air and outdoor air. The colder the outdoor air, the lower the proportion of outdoor air to return air required. The three damper motors operate together, the outdoor air and relief air damper motors moving in one direction and the return air damper motor moving in the opposite direction.

As the outdoor air temperature rises, a larger and larger volume of this air is required to produce the desired effect. At some point the temperature of the outdoor air is adding to the cooling load of the building rather than providing a free cooling effect. To prevent

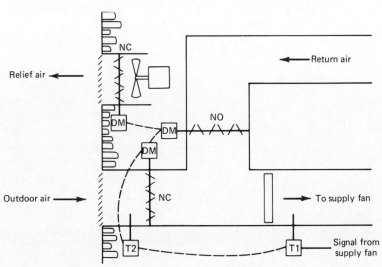

FIGURE 6–8   In this economy cycle configuration, the relief fan could be omitted and the relief dampers could tap directly into the return ductwork. Note that this system is only operable when the supply fan is energized.

this, thermostat T2 is used to override the signal of T1 if the outdoor air temperature rises too high, say above 65°F. When T2, referred to as a high limit, senses too high a temperature, it will cause the outdoor air damper and relief damper to return to a minimum position so that the return air damper will open.

In Figure 6–8 the design provides for a minimum outdoor air damper position. In this way, even though the economizer is not in operation, the required amount of ventilation air is still getting into the building. In Figure 6–11 the minimum ventilation air requirement is handled by a separate damper and damper motor tied into the supply fan. The economizer operates independently and will shut off completely when not required.

A variation of this system uses an enthalpy control in place of the high limit thermostat T2. It is conceivable that air at 60 to 65°F would not provide a cooling effect if it were high humidity air. Although it might provide sensible cooling, it would add to the latent load of the building and the net effect would be to decrease rather than increase the cooling ability of the system. The enthalpy control measures the total heat content of the outdoor air and only permits the economizer to operate if the enthalpy is low enough to produce a cooling effect.

FIGURE 6–9   In smaller systems incorporating an economizer cycle, the control package is quite simple. The damper motor controlling both outdoor air and return air dampers is controlled by the package shown here. Sensors for outdoor air and mixed air control the position of the dampers. (Courtesy of Honeywell Inc.)

TWO-DAMPER LINKAGE

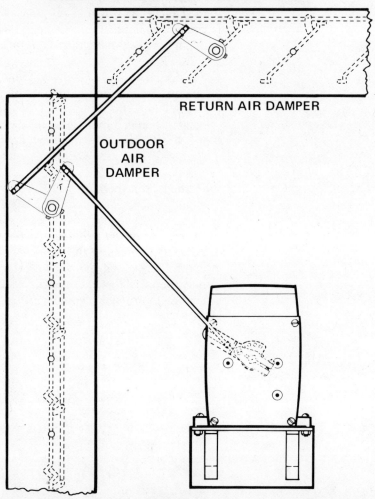

RETURN AIR DAMPER

OUTDOOR
AIR
DAMPER

FIGURE 6–10   In smaller systems it is not uncommon to have a single motor drive both the outdoor air and return air dampers if the return air ductwork and outdoor air ductwork are sufficiently close together. (Courtesy of Honeywell Inc.)

6.4.3 Air Cleaning   The importance of clean air, that is, air free of gaseous and particulate contaminants, cannot be overemphasized. Much discomfort and illness with accompanying absenteeism and lost production time can be specifically attributed to respiratory problems caused by airborne contaminants. The fact is that air cleaning is a sophisticated science that promises great return for mankind in the future.

Let us consider the *typical* central air conditioning system that exists today in most small and large installations. The two main

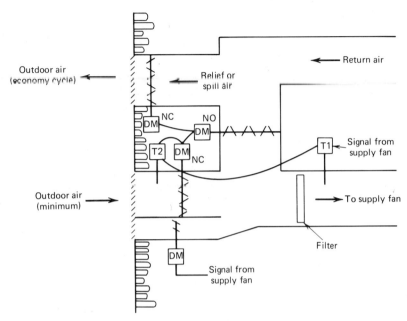

FIGURE 6–11   A minimum ventilation air damper of the parallel blade type, commonly used when precise metering is not required. The opposed blade type, shown in the return air duct, spill air duct, and economy cycle outdoor air duct provide better control characteristics.

reasons for any kind of air cleaning equipment in such systems is (1) to prevent particulates from clogging the air passages and reducing heat transfer efficiency, and (2) to remove heavy particulates from the air that might otherwise accumulate in the conditioned space in the form of unsightly deposits.

Much air cleaning equipment has been developed and installed that will remove smoke and odors from restaurants and other public places and that will remove bacteria and viruses from air going into certain hospital areas. The technology exists and is being applied today, but the vast majority of installations are still using minimum filtration for the reasons already mentioned.

It is a shame that even in the areas where minimum filtration is used, it is neglected as often as not. Filters should be cleaned or changed on a regular basis, since they tend to inhibit airflow significantly as they gather dirt, causing reduced cooling or heating capacity. In extreme cases in direct expansion cooling units, blocking the airflow causes the liquid refrigerant to slug the compressor and sometimes creates low pressures and temperatures with resultant coil freezeup. Removing the filters permanently, which is often done by uninterested or uninformed maintenance people or equipment owners, will cause clogging of the heat transfer surfaces with the same result.

FIGURE 6–12   Three sets of pneumatically controlled dampers. At the top the return air dampers are fully open permitting return air to drop down into the return air plenum. The open set of dampers is for minimum outdoor air requirements. Immediately below these dampers is the third set, closed, which is actuated by an unseen pneumatic actuator in response to the demands of the economy cycle controls and introduces outside air for free cooling.

The most commonly used filters are the throwaway fiberglass types, which come in a 1-inch thickness for small residential equipment and a 2-inch thickness for larger commercial equipment. They are also available in thicker sizes for computer rooms and other specialized equipment. They must be replaced periodically. Usually, a simple differential pressure indicator is used in larger units to indicate when the restriction to flow is becoming excessive. Figure 6–15 depicts one such indicator.

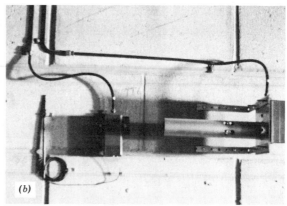

FIGURE 6–13   The enthalpy sensor mounted in the outdoor air duct (a) is used to provide a signal that is compared to the enthalpy sensed by an enthalpy sensor in the return air duct (b). If the outdoor enthalpy is significantly lower than the indoors, outdoor air will be used for free cooling. Tubing indicates a pneumatic system.

To minimize the number of times filters have to be changed and the labor associated with such changes, and to insure that clean filters are always available, the roll-type filter is used in some units. The roll filter can be manually advanced periodically as it becomes dirty, but in its more sophisticated form it is advanced automati-

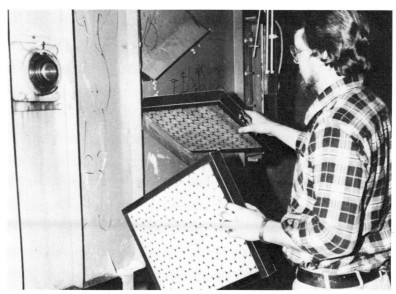

FIGURE 6–14   Dirty 2-inch throwaway fiberglass filters are being replaced in this airhandling unit. The frequency of replacement will vary between installations dependent upon air quality, the volume of outside air being used, and the cleanliness level of the building.

FIGURE 6–15   This filter change indicator measures the differential pressure across the filter. When this pressure reaches an unacceptable level, as suggested by the filter manufacturer's instructions, the "change filter" flag drops indicating that it is time to change the filter. A built-in electric switch can be used to set off an audible or light alarm. (Courtesy of Honeywell Inc.)

cally. A motorized mechanism may be geared to continually move the filter media at a very slow rate. In another scheme the pressure differential across the filter is monitored and when it becomes excessive, a motor automatically advances the roll to bring fresh media into the airstream.

Electronic air cleaners have been growing in popularity primarily because of their ability to remove smoke from the air. Although fibrous media filters tend to become more efficient as they become dirty, until they are so efficient they do not permit even air to pass through, the electronic air cleaner tends to become less efficient. As dirt accumulates on the collector plates, the efficiency drops requiring that the plates be periodically cleaned. The frequency of cleaning, of course, depends upon the location and air quality. There are control systems that automatically activate a power spray cleaning mechanism (after deenergizing the air cleaner) on a time schedule. The complexity of such systems compared to the throwaway filter is one important reason why they have not been growing in use as rapidly as one might think.

6.4.4 Humidification   Dehumidification is just as important to comfort in summer as the addition of moisture to the air in winter. Cold outside air entering a building through infiltration and as ventilation air will cause the relative humidity to drop. It is not unusual to see indoor relative humidity levels as low as 10 to 15%. Under such conditions the normal defense mechanisms of people dry out and they become susceptible to respiratory illness. Furniture dries out and warps, and static electricity builds up causing discomfort and also causing problems with some industrial processes. It also seriously affects computer operation. As a general rule, relative humidity levels much below 35% are unsatisfactory.

Moisture is added to air in a number of ways ranging from the ingenious to the amazingly simple. Air passing through a water spray will pick up moisture. In some systems the spray may be quite heavy in which case the air is cleaned as well as humidified. The dirt from the air runs off with that portion of the water that does not evaporate. In other systems the water is atomized and introduced to the airstream at a rate that prevents the carry-over of liquid into the conditioned space. If the air has been heated first, the evaporation process is much more rapid; in many applications the humidifying apparatus is installed downstream of the heating coil.

Another technique employs a bath of water in the airstream. Submerged in the bath is an electric heater. The amount of water vapor introduced to the air is directly related to the amount of heat added to the water in the bath. Rather close control can be obtained with such a system.

A very popular arrangement in small residential humidifiers uses a porous pad through which air flows. In some schemes the water flows over the pad whereas in others the pad, in cylindrical form, rotates, dipping into the water bath and carrying the moisture into the airstream.

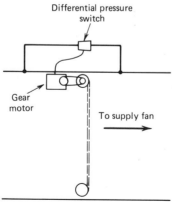

FIGURE 6–16   The differential pressure switch, sensing that the filter media is becoming clogged, sends a signal to the gear motor to advance the media roll. The motor has a timer mechanism built in that enables it to advance the roll enough to bring a fresh media pad into the airstream, and then it stops to await another signal from the pressure switch.

FIGURE 6–17   An electronic air cleaner available in efficiencies as high as 95%. The key to efficient operation and long life for this device is maintenance. It must be cleaned regularly. The need for cleaning is signaled by the lighted switch on the control panel at the top of the unit. (Courtesy of Honeywell Inc.)

Perhaps the most obvious humidifying technique is to use steam. Obvious, that is, if steam is available in the building. If not, steam generators designed specifically for humidifying applications can be purchased. Such a steam generator would only be economically sound in a building of some size.

All humidification equipment has one major drawback: They all use water. Actually, water (H_2O) itself is not too bad; the problems arise because of the dirt and minerals dissolved in the water. In every method described here the water vaporizes and leaves the

OUTDOOR-INDOOR RELATIVE HUMIDITY CONVERSION CHART

OUTDOOR RELATIVE HUMIDITY (%)											
100	2	3	4	6	7	9	11	14	17	21	26
90	2	2	4	5	6	8	10	12	15	19	23
80	2	2	4	5	6	7	9	11	14	17	20
70	1	2	3	4	5	6	8	10	12	15	18
60	1	2	3	3	4	5	7	8	10	13	15
50	1	1	2	3	4	4	6	7	9	10	13
40	1	1	2	2	3	4	4	6	7	8	10
30	1	1	1	2	2	3	3	4	5	6	8
20	–	1	1	1	1	2	2	3	3	4	5
10	–	–	–	1	1	1	1	1	2	2	3
0	0	0	0	0	0	0	0	0	0	0	0
	-20	-10	-5	0	+5	+10	+15	+20	+25	+30	+35

OUTDOOR TEMPERATURE (°)

Outside Temperature	Recommended Humidity
+ 20 and above	35%
+ 10	30%
0	25%
− 10	20%
− 20	15%

FIGURE 6–18   The relative humidity that can occur indoors when outdoor air at a temperature and relative humidity as indicated on the chart is introduced into the building is shown at the left. The chart at the right shows the desirable relative humidity to be maintained indoors at indicated outdoor temperatures. It is clear from these two charts that humidification is a desirable air conditioning system capability in cold winter climates.

minerals behind. These minerals must be removed or eventually they jam the equipment, cause electric heaters to burn out, cake up in the duct (in the case of the atomizer type) and, in general, cause problems.

Anyone with a humidifier or anyone who is planning to install one should be aware that maintenance of this equipment is no less important than the maintenance of the other parts of the HVAC system.

Control of humidification equipment is rather simple. A humidistat has an element sensitive to moisture content. It can be located in the return air duct or in almost any place in the conditioned space. Unlike a thermostat that must be located in a spot sensing the average temperature of a space, the humidistat can be located practically anywhere. Humidity differences in the space are minimal as stratification and humidity "deadspots" cannot occur. Water vapor in a space will distribute itself in a manner predicted by Daltons law of partial pressures. In practice, although not 100% correct, it is close enough to being true to believe.

One final word on humidification. Humidifiers work best in ducted systems; however, humidification equipment is available for piped systems as well. Of course, the steam humidifier is ideal; a little steam is injected into the air on the demand of a humidistat and that is all there is to it. Other systems using atomizers or self-contained units with wetted pads and fans have also been developed. A pan-type humidifier has even been developed that fits inside a residential baseboard radiator. Humidification is available for any application.

6.4.5 Static Pressure Control

In a number of air conditioning applications it is desirable that the pressure in one space be maintained at a different level than in an adjoining space. As an example the kitchen area of a restaurant might be at a lower level of pressure than the dining area so that cooking odors will be contained. An isolation ward of a hospital may be kept at a lower pressure than its surroundings so that contagious diseases are not spread. A "clean room," that is, a facility for assembling delicate mechanisms under dust-free conditions, will be kept at a positive pressure relative to its surroundings. Stairwells in some buildings may be pressurized in the event of a fire to prevent smoke from entering and slowing down evacuation. These are some common examples—with some additional thought the list could be expanded considerably.

Static pressure control is achieved either by introducing more air to a space than is being exhausted in order to maintain positive pressure, or by exhausting more air than is being introduced to maintain negative pressures. This is most commonly done by manipulating supply and exhaust dampers using a static pressure

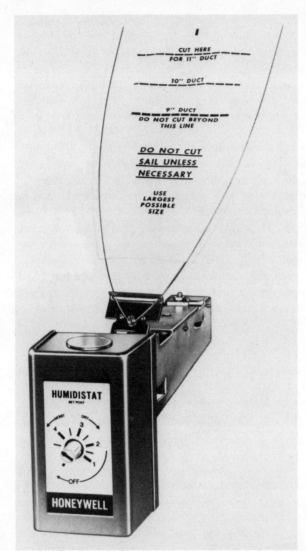

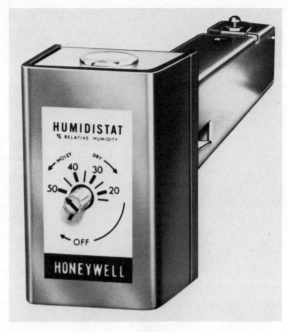

FIGURE 6–19   Two humidistats. The control at the left incorporates a sail switch which insures that the humidifier will not operate unless the hot air system is in operation. The control at the right must be electrically interlocked with the furnace fan to provide this feature. The element of both controls is inserted into the airstream to sense the relative humidity. The set point is adjustable by means of the knob on the face of the control. (Courtesy of Honeywell Inc.)

controller. This device senses two pressures, the reference pressure and the controlled pressure. In the case of the restaurant example the pressure in the kitchen could have been the controlled pressure and the dining area pressure could be the reference pressure. The kitchen exhaust system would be equipped with

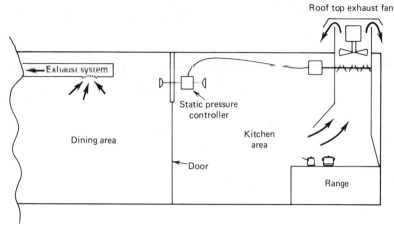

FIGURE 6–20   If the dining room pressure drops slightly because of too much air being exhausted, the static pressure controller adjusts the exhaust damper in the kitchen to increase its exhaust air volume. The negative pressure in the kitchen relative to the dining room is then maintained.

dampers that respond to the static pressure controller. The pressure differential maintained is typically on the order of 0.01 to 0.02 inches of water column, not a very large differential at all, yet enough to insure airflow from the dining room into the kitchen and not the reverse.

The same effect could be achieved by reversing the process and having the kitchen as the reference and the dining area as the controlled space. In that case the outdoor air dampers of the dining room HVAC system would be controlled by the static pressure controller with more air being introduced to the dining area relative to the kitchen to maintain the desired differential.

In any situation where a pressure differential is desired, we should be aware that too much of a differential is as bad as too little. A high pressure difference causes air noises as air flows through all the cracks and crevices between the two spaces. Also doors between the spaces may be harder to open, and once opened, harder to close. Swinging doors may be continually ajar. Note also that there is an energy cost involved in maintaining a differential, a cost that is particularly high when the differential is between an indoor and outdoor space. Pressure differential, then, should be as low as possible and applied only in those situations that warrant it.

6.4.6 Energy Reclamation   A major waste of energy occurs whenever warm or cool air is exhausted from a building so that ventilation air can be introduced. Devices have been around for quite a while that attempt to reduce this waste. Actually, energy reclamation is not new—it has just been given a lot of publicity recently and the absolute dollar amounts being wasted are becoming increasingly significant. There are many areas of energy waste in a building, but in the case of "designed waste" such as we have in an exhaust system, corrective action can be designed as well.

The regenerative wheel shown in Figure 6–21 has been used for a number of years. Some early designs were installed in stacks to reclaim a portion of the heat going up the chimney in large central heating plants. Improved designs with efficiencies of up to 70% have appeared in exhaust air systems in recent years. Any efficiency figures assume well-maintained, clean equipment, but this is not necessarily the case. Warm air exhausted from the building heats the wheel media, which rotates into the stream of cool incoming air, where it gives up heat. The entering outside air is therefore preheated. This is not totally free heat, since the pressure drop through the wheel must be overcome by increased fan horsepower and additional maintenance costs must be included as well as the not insignificant cost of the equipment itself. Energy savings are certainly possible, but careful study is required before any energy reclamation system is installed.

The heat pipe shown in Figure 6–22 has a closed refrigeration system that recovers waste heat. The refrigerant in the section of the pipe in the warm airstream vaporizes, absorbing heat, and flows to the section of the pipe in the incoming cool airstream, where it condenses releasing heat. The condensed refrigerant will flow by gravity to the warm section and repeat the cycle. This recirculation will continue as long as there is a temperature difference between incoming and outgoing air. By assembling a number of heat pipes in the form of a coil, a considerable heat transfer capacity can be developed. Both the wheel and the heat pipe will work in winter, when the exhaust air is warm, and in summer when the exhaust air is cool. Depending on the amount of ventilation air involved, the size of the heating and cooling systems can be substantially reduced if these devices are incorporated in new designs.

Both the heat pipe and the wheel require that exhaust and ventilation duct work be brought together to pass through the heat reclaim device. In some installations this could pose a real design problem. The "run-a-round" system provides one solution. A standard fin tube heat exchanger is put in both ducts, no matter how far apart. A pump circulating water through both coils will achieve a transfer of heat from the warmer to the cooler coil. Although proponents of each of these systems claim heat recoveries bordering on the miraculous, it is not unreasonable to expect properly installed and continually maintained systems to produce efficiencies on the order of 60%.

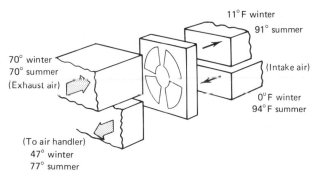

FIGURE 6–21 Notice the ability of the regenerative wheel shown here to increase the incoming outdoor air temperature in winter and decrease the incoming air temperature in summer. The ventilation air must then be further tempered by the main air conditioning system, but energy savings can be readily identified with this arrangement.

The recovery of heat being wasted through exhaust air is only one area for savings. Heat pumping systems that shift heat from parts of a building where it is not needed to areas where it can be used will be covered in the next chapter.

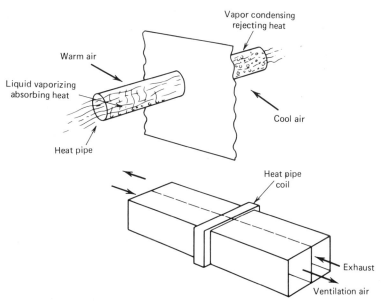

FIGURE 6–22 The single heat pipe has the ability to transfer heat from one airstream to another. When ganged into a coil arrangement, an effective energy conservation device is obtained. Note that the two streams of air must be brought into close proximity of one another.

6.4.7 Fire Safety

Increasing attention is being paid to the idea that a building is a total system rather than a collection of independent subsystems. Fire safety is becoming an important part of the building design process, and a growing number of design firms have specialists in this area. Because a number of fire safety considerations interact with HVAC considerations, it is important that the HVAC controls person be aware of them.

The greatest hazard to life in a fire is the smoke generated. Such smoke inhibits evacuation of a building by obscuring safe routes of exit, creates panic in the minds of the occupants, and can overcome people causing injury or death. It is the purpose of the fire safety system to detect the existence of a fire as early as possible, to sound a warning, and to extinguish the fire or at least contain the fire and its effects until fire fighters can bring it under control.

Detection can be done by heat sensors, visible light sensors, smoke detectors, and more recently by ionization detectors that sense the products of combustion at a very early stage in the development of the fire. Since the air conditioning system circulates air, it makes sense to provide the system with some of the fire safety functions.

Ducts that pass through fire-rated walls, that is, walls designed to contain the spread of the fire, must be equipped with fire dampers. These dampers are typically spring-loaded and have fusible links. When the temperature in the duct exceeds a safe level, the link will melt causing the damper to close and seal off the duct. The intent is to prevent flame and smoke from spreading throughout the building by means of the ductwork.

Another commonly used control is a thermostat, referred to as a *firestat*, in the return air ductwork. It is set to shut down the air conditioning system when temperatures rise to about 125°F. At that point it is apparent that a fire is present and, if the system continues in operation, it will spread fire and smoke throughout the building. In some systems an optical sensor will be used to sense the presence of smoke, since such smoke may very well be present even though temperatures at the firestat are considerably lower than 125°F. Ionization devices may also be used in some newer designs.

Although the most common techniques in use today call for the air conditioning system to be shut down in the event of fire, other techniques have been used from time to time, particularly in recent years. One such system calls for a series of dampers to shift position causing the building air to be dumped outside and 100% outdoor air to be brought into the building. The outdoor air may have the effect of feeding oxygen to the flames, but the dumping of the indoor air has the effect of lowering the smoke levels giving the occupants a better chance to escape.

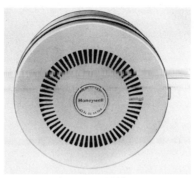

FIGURE 6–23   Ionization-type smoke detectors can signal the existence of a fire before visible indications can be seen. They are widely used in commercial and residential applications and are available to provide audible and/or visual warning signals. (Courtesy of Honeywell Inc.)

Another scheme calls for the pressurizing of stairwells relative to the building proper. Smoke cannot infiltrate into the stairwells and they will remain clear, enhancing the chances for escape. Commonly, the stairwell system is totally independent of the main HVAC system and is activated only when a fire is detected. Still another technique calls for floors or zones above and below the fire floor to be pressurized. The effect is to contain the smoke on the fire floor.

The National Fire Protection Association (NFPA) generates standards covering all aspects of fire control including the standards to be used in air conditioning systems. Typically, local building codes will use these standards as reference by requiring that systems installed in the local jurisdiction meet all or part of these standards. It is the function of the building department to review plans and make sure that the requirements of the building code are met. This same department also employs inspectors who must physically inspect the building as it is going up and then finally after it is completed, to ensure that the plans have been followed and that the building meets code requirements. At that time the department issues a CO (Certificate of Occupancy), the document at the end of the rainbow, which usually signals that the job is done and the building is ready to be used.

DISCUSSION TOPICS

1. **Describe what is meant by the core and perimeter of a building.**
2. **Why is ventilation in a building required? List several different applications and describe the considerations that might determine ventilation air requirements.**
3. **What is the purpose of a relief fan?**

4. How does an economy cycle provide free fooling? Is it really free?
5. What humidity would you expect indoors on a day when the outdoor temperature is 20°F and the relative humidity is 70%? Is this too low?
6. Why is a high pressure differential between indoors and outdoors undesirable in a building?
7. Where might a regenerative wheel prove effective?
8. How does the HVAC system play a role in fire safety?

Matching Building Requirements

Once the building function has been studied and the requirements of the HVAC system identified, it is necessary to select the component subsystems that, when assembled, will meet the requirements. In this chapter we shall discuss the systems that are in common use today in large installations.

7.1 DUCTED SYSTEMS—SINGLE ZONE

The simplest ducted system that provides heating or cooling is shown in Figure 7–1. It is a single-zone system designed to provide either heating or cooling but not both at the same time; it also has provision for a fixed amount of outside air. The heating is provided by a duct coil connected to either a hot water or a steam boiler. In smaller systems the heat may be provided by a direct-fired system using gas or oil or even electric heaters.

Control may be achieved by a thermostat, which may be either two-position or proportional and located in the return duct as shown in Figure 7–1. It could also be located in the conditioned space. A more sophisticated arrangement uses a high limit to override the signal from the main thermostat in the event that the air temperature in the duct is adequate for heating or cooling. This system can also be used with a master–submaster arrangement with the duct temperature being reset by the thermostat in the conditioned space.

Minimum air is provided by outdoor dampers controlled by the main supply fan circuit. When the supply fan is energized, the dampers open to admit ventilation air.

The cooling function is provided by a separate coil in series with the heating coil. It may be a DX coil connected to a condensing unit or it may be a chilled water coil. Control of the cooling function is usually accomplished by the same thermostat as that handling heating. Although it is possible to use a thermostat that calls for heating or cooling with a dead band between each mode, it

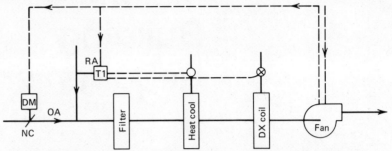

FIGURE 7–1   The single-line method of representing control configurations is very commonly used in the field. In this simple single-zone system T1 controls both heating and cooling. The dotted line indicates that the control system is only in operation if the supply fan is working. The outdoor air damper is of the two-position type also being energized when the supply fan is on.

is more common to change over from one mode to the other manually. In its simplest form this system will have either hot water or chilled water available depending on the time of year or outdoor temperature so a thermostat with an automatic change-over capability is usually not warranted.

Although this system is a single-zone unit, a building could very well be divided into a number of zones with each zone being serviced by a complete system as shown in Figure 7–1. It is quite common, in fact, to have a multistory building zoned in exactly this way with a number of floors being handled by a single unit. In effect a 20-story building is actually four 5-story buildings stacked on top of one another.

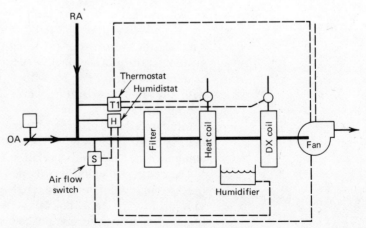

FIGURE 7–2   In this single-zone system the humidifier will only operate if the airflow switch senses that the fan is actually moving air through the ductwork.

In Figure 7–2 a pan-type humidifier has been added. It has been placed after the heating coil because humidification is usually required in winter and air that has been heated will absorb more moisture than cool air. Note the humidistat located in the return air duct. It will energize the electric heater located in the pan when humidification is required.

There are two schools of thought on whether the heating coil or cooling coil should be closest to the incoming air. If very cold outdoor air is expected, having the heating coil act first as a preheater would be useful. If dehumidification in the summer is a concern, placing the heating coil after the cooling coil, to act as a reheater, would be a sound approach. If both requirements must be met, a separate preheat coil might have to be used. In fact, it is not uncommon to have a small preheat coil situated in the path of the outdoor air only, to bring it up to a usable temperature, say about 55°F.

In Figure 7–3 the basic single-zone unit has had an economizer added. The room thermostat will energize the economizer on a demand for cooling. Thermostat T1 causes the outdoor air and return air dampers to modulate in an effort to provide sufficient cooling when using cool outside air to satisfy the room thermostat. If this is not adequate, the mechanical refrigeration system will be energized by a second stage of the room thermostat in order to add enough cooling to do the job. If the outdoor air is too warm to do *any* cooling, T2 will override T1 putting the dampers in the minimum outdoor air mode and will immediately signal the mechanical refrigeration to assume the entire load. There are a number of variations in this sequence of operations, but they all have the same purpose, that is, maximizing cooling by minimizing energy consumption.

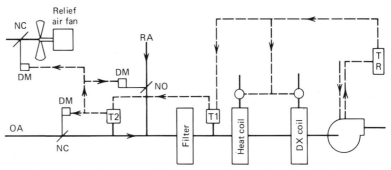

FIGURE 7–3   In this single-zone system with an economizer-cycle, thermostat TR has a two-stage cooling capability. The first stage energizes the economizer cycle. If that is not adequate, continuing rising temperature causes the second stage to energize the refrigeration system connected to the DX coil.

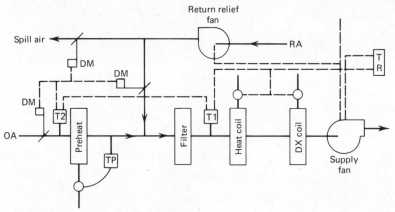

FIGURE 7–4   In this single-zone system a return relief fan has been installed to insure that adequate air is exhausted from the building. This fan also is designed to overcome return duct resistance insuring that the air pressure in the vicinity of the filter bank is very close to atmospheric pressure. Note the thermostat TP, which will preheat the outdoor air used for ventilation in winter by controlling the flow of hot water to the preheat coil.

Note that in Figure 7–3 a relief fan has been installed. Since larger amounts of outdoor air are required in an economizer system, a technique for exhausting air must also be included. Figure 7–4 shows a somewhat similar arrangement. The main difference is that the relief fan shown has a second purpose, to overcome the resistance of the return ductwork. It is sometimes called a return air fan or return-relief fan. The dampers are arranged so that varying amounts of air can be exhausted from the building based on the quantity of outdoor air brought in. Usually, the amount exhausted is about 10% less than that brought in so that a slight positive pressure can be maintained in the building.

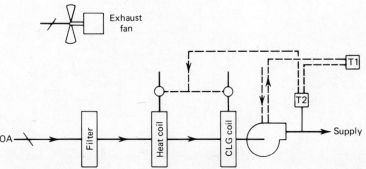

FIGURE 7–5   In this 100% outside air system the dampers are of the gravity type that open when the fans are energized and automatically close when they are idle. The control configuration shown has the zone thermostat resetting the set point of T2, which in turn controls the heating/cooling function.

Figure 7–5 depicts a 100% outside air system. It is used in applications where recirculating indoor air is not a good idea. Some industrial processes produce fumes that should be exhausted from the building. In certain medical facilities, recirculated air is undesirable. The fresh air being brought in is sometimes referred to as make-up air. The system shown has a preheat coil controlled by duct thermostat T1 with a set point of 55°F. The valve supplying hot water to the coil is set so that a small flow of hot water is always passing through the coil. In the event of a control failure there will be sufficient flow to prevent a coil freezeup. Note the filters in all these systems. They are vital and must be maintained. This is particularly true in those systems employing more than minimum outside air, where the filter life decreases as outdoor air pollution increases.

7.2 DUCTED SYSTEMS—MULTIZONE

Using multiples of a single-zone unit in a building with a large number of zones is not particularly economical. A single unit with the ability to provide heating or cooling on demand to a number of zones year round would be more desirable. Such systems are called multizone systems.

Figure 7–6 shows the one system that usually comes to mind when the expression "multizone unit" is used. It incorporates a heating and cooling capability in a single piece of hardware. The cooling coil provides cold air through the "cold deck" and the heating coil provides hot air through the "hot deck." As mentioned

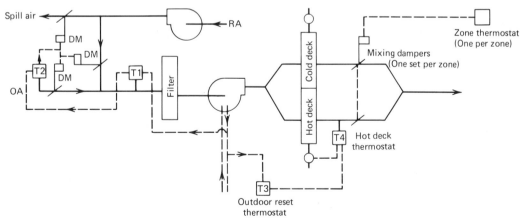

FIGURE 7–6   In this multizone system it is desirable to prevent simultaneous operation of the boiler and refrigeration systems. Thermostat T3 cannot only reset the set point of the hot deck thermostat T4, but also can be used to energize the cooling system. Alternatively, DX equipment may have outside-mounted thermostats to prevent operation until outdoor temperatures reach a predetermined level.

previously, any heating or cooling technique can be incorporated in such an arrangement.

Varying quantities of hot and cold air are combined by the mixing dampers and sent to the zone. The ratio of hot to cold air is controlled by the zone thermostat. There is a set of mixing dampers and a thermostat for each zone. It is not unusual to have a single multizone unit with, for example, seven sets of mixing dampers leading into seven ducts that carry air to seven zones. The air temperature in each duct could be different depending on the demand sensed by each zone thermostat. The volume of air flowing in each zone duct does not change; however, the ratio of air from the hot deck to air from the cold deck does change in response to the zone thermostat.

Prior to the energy crisis it was common to see systems that had the boiler and compressor running year round. In effect, the boiler was used to heat air that was simultaneously being cooled by the compressor system. Newer energy codes specifically prohibit that. In midwinter the heated and return air are combined whereas in midsummer the cooled and return air are combined. If extra cooling is required during in-between seasons, an economizer is employed.

FIGURE 7-7   A view of the cold deck mixing dampers of a multizone system handling a 60,000-cfm airflow. Note that five sets of dampers are shown. The airflow is from left to right. Immediately above these dampers is a similar arrangement of hot air mixing dampers. The motors operating these dampers are located below, behind the sheetmetal partition. Note the access door at the upper left permitting access to the hot deck dampers.

FIGURE 7–8   A view looking downstream through the cold deck mixing dampers. Note the turning vanes several feet down the duct. This zone is calling for full cooling. The zone served by the dampers to the right is calling for almost full heat, which in this system at this time of year (summer) means full return air and no cooled air. The air in the duct beyond the dampers is a mixture of cool air passing through the dampers shown and return air flow through the hot air dampers immediately above. At the lower right, barely seen, is an access door to an area containing the damper motors.

1. Midwinter. In Figure 7–9 the requirement in Zone 1 is for heating so that the mixing dampers are more open to the hot deck with its hot air than to the cold deck through which is passing unconditioned return air. In Zone 2 the thermostat is satisfied and the balance of hot deck air and cold deck air is such as to maintain that condition. This does not necessarily mean a 50:50 ratio of hot to cool air. In Zone 3 there is a requirement for cooling that will be met by recirculating the untreated air (as well as the cold outdoor ventilating air) passing through the cold deck.

2.   Midsummer. In this case the boiler is shut down and the refrigeration system is in operation. The cold deck temperature is low due to the action of the compressor system and the hot deck temperature is warm due to recirculating unconditioned air.

3.   In-between seasons. The boiler is on to provide heating and the economizer is activated to provide cooling. The boiler output can be tempered by using an outdoor air thermostat to reset the hot deck temperature. A common setting is to maintain a 100°F hot deck temperature at a 70°F outdoor temperature. A reset ratio of 1:1 will result in the hot deck's increasing to 120°F when the

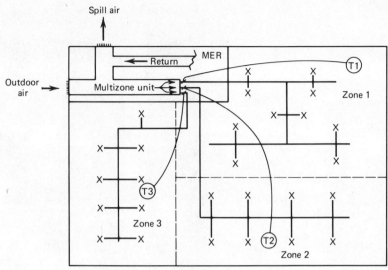

FIGURE 7–9   A multizone unit located in the mechanical equipment room (MER) with three sets of zone dampers (one hot deck and one cold deck damper per set) each controlled by a zone thermostat. The set point of T1, T2, and T3 can each be different, as each thermostat controls the temperature of the air in its zone independent of the other zones.

outdoors drops to 50°F. The temperature at which the boiler should start up and the mechanical system should shut down in favor of the economizer will vary from building to building. It is up to the *building operating personnel* to determine this point using the design engineer's prediction as a starting point.

A close relative of the multizone unit is the dual duct system. In this type of system, mixing does not occur at the unit but rather close to the zone. Two ducts carry air, one hot and one cold, to mixing boxes at the zone. The ratio of hot to cold air is controlled by the zone thermostat through the mixing box.

Figure 7–10 shows a mixing box for such a system. The volume of air entering each zone is constant while the temperature is varied. The initial cost of such a system is higher than the standard multizone system because of the extra duct work required as well as the zone mixing boxes. Response to load changes can be quite rapid, however, since mixing does occur at the zone.

It is common for a central air handler to provide hot and cold air to dozens of mixing boxes throughout a building. Each mixing box can then be connected to as many diffusers as necessary to serve the conditioned space. This is an advantage over the multizone unit, which is somewhat limited in the number of zones it can serve, with perhaps about a dozen being the upper limit.

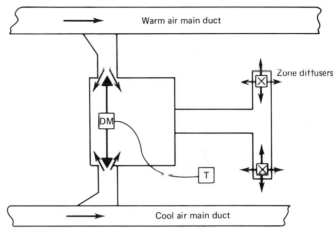

FIGURE 7–10   The dual duct system has been used increasingly in retrofit work, since it can turn an existing "heating only" system into a "heating/cooling" system rather economically. The mixing box can provide conditioned air to any number of diffusers in a zone.

Several years ago dual duct systems using high velocity air were quite popular. The need for sound attenuation (deadening) in the mixing boxes as well as the necessity for rather tight sheet-metal joints to minimize noise has led to its falling into disfavor.

In recent years the variable volume system has been growing in popularity. Air at a fixed temperature is sent to each zone. The zone thermostat modulates a damper, sometimes in the duct and sometimes located in the diffuser itself, to vary the volume of air entering the space. Such an arrangement has been found to be economical because of the energy savings in moving only the required amount of air through the system rather than a constant volume at all times. Most systems can effectively distribute air to the conditioned space at system volumetric flow rates as low as 50% of the full-rated flow. The air horsepower or brake horsepower or energy required is reduced by a similar amount, which over a year can amount to a healthy dollar savings.

The temperature of the air in the VAV system is commonly controlled by a master–submaster arrangement of thermostats related to the outdoor temperature. Although cooling and heating are not simultaneously available year round, the system can respond to a fairly wide swing in outdoor temperatures and can maintain more comfortable temperatures than a single-zone system.

Still wider outdoor temperature excursions can be accommodated with the addition of reheat to the zones. The main supply temperature can then be maintained at a lower level for those who enjoy a cool environment, and extra heat is available through the reheat coils for those who want it. Typically, the thermostats will

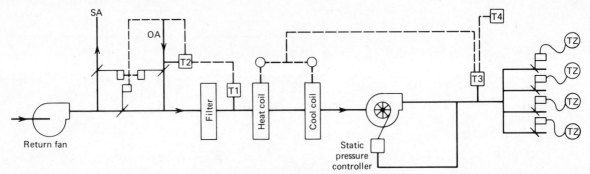

FIGURE 7–11   In this VAV system the zone thermostat TZ will vary the airflow to each zone by controlling a damper in the VAV terminal. As the airflow decreases, the pressure in the main supply duct will tend to increase; however, the static pressure controller sensing this sends a signal to the inlet vanes of the supply fan and causes them to restrict the airflow to the fan and maintain a constant duct static pressure.

gradually increase the flow of air to the zone as demand increases and will then energize the heating coil to maintain temperature at higher heating demands.

A characteristic of VAV thermostats of interest is the requirement for changeover from heating to cooling. In winter, when more heat is needed, the thermostat must cause the damper to open. In summer, when less heat is needed (or more cooling is required), the thermostat must also cause the damper to open. The signal generated by the thermostat must be reversed from one season to the next whether it is accomplished automatically or manually.

To avoid high air pressures in VAV duct systems, pressure sensors are used. They will sense the increasing pressure as a number of VAV terminals close down and relay the signal back to the main supply fan. Inlet vane dampers will then throttle down decreasing the capacity of the fan with an accompanying drop in power consumption. As the VAV terminals open, the duct pressure decreases causing the sensors to send a signal to the supply fan inlet vane dampers making them open again.

Ducted systems capable of serving a number of zones in a building year round have proved to be quite versatile. They are limited, however, to commercial buildings, public buildings, and other nonresidential applications. Additionally, they have been found to be not quite as effective as one might wish in perimeter areas of buildings. To fill these gaps, we have piped systems.

7.3 FAN COIL SYSTEMS

As described previously, the name *fan coil* describes these systems perfectly. A fan blows air across a coil, through which flows hot water in winter and cold water in summer. The terms *two-pipe, three-pipe* and *four-pipe* fan coil describe systems of increasing sophistication and cost.

The two-pipe fan coil has a supply line carrying chilled or hot water to the terminal and a return line carrying the water back to the central plant; hence the term *two-pipe*. The temperature of the water supplied is determined by the central control system and rarely changes much from day to day. Increasingly, controls are being applied that will provide the lowest-temperature water that will still be warm enough to heat the building in winter and the highest-temperature water that will still cool the building in summer. Although changeover from cooling to heating may be automatic based on outdoor temperature, it is more frequently manual. Depending on the locale, magic dates are written into tenant leases indicating that after May 15th only cooling will be available and after October 15th only heating will be available. This, of course, is a reflection of the operating personnel's practice of manually changing over the system on those dates. Of course, if one needs heat in June or cooling in November, one is out of luck, but for many it has proved to be a perfectly satisfactory method of operation.

The three-pipe fan coil is a step up in comfort. It is intended to provide heating and cooling year round. There are two supply lines going to each unit: One carries hot water, the other cold. Again one return line is used, and herein lies the problem. Hot water at 160°F passes through the fan coil in an apartment requiring heat

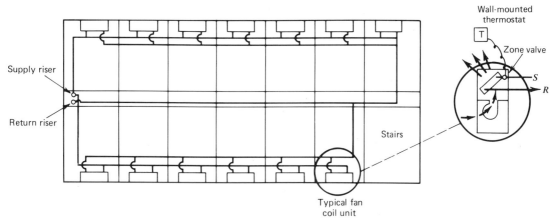

FIGURE 7–12   The two-pipe fan coil system in this planview of a motel is the most economical in operation and in initial cost but the least able to cope with varying outdoor temperature. Although in theory the temperature of the water can be changed during the course of the day to provide heating in the morning and possibly cooling as the day wears on, in practice this is usually not done due to the thermal inertia of the system.

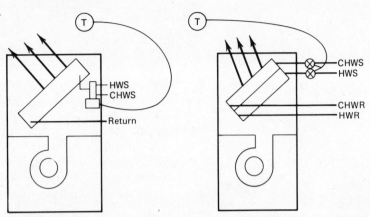

FIGURE 7–13   In the three-pipe fan coil unit shown at the left, either hot or chilled water flows through the coil upon demand of the thermostat. At the right a split coil is used for the four-pipe configuration. The thermostat again determines whether heating or cooling is required. In some designs a single coil is used for both heating and cooling. In such cases two three-way valves are used, one on the supply and one on the return to direct the water in the appropriate manner.

and passes into the return line at 140°F. Chilled water enters the fan coil of an apartment requiring cooling at 45°F and passes into the same return line at 55°F. The mixing of the two streams of water results in a mixture at 95°F. This water then returns to the central plant, where that water passing through the chiller must be cooled to 45°F and that passing through the boiler must be heated to 160°F. It ought to be plain that much energy is being expended in this very wasteful process. Today three-pipe fan coils are practically extinct and those remaining in the field are ripe for conversion to another system, perhaps four-pipe. Incidentally, in midsummer or in the dead of winter, when most if not all of the fan coils are in the cooling or heating mode, the system is fine. It is the in-between seasons that are disastrous. The problems of the three-pipe system are solved by the addition of a second return line, making a four-pipe system. The heating piping and the cooling piping are now totally separated. On a call for heat, hot water flows through the coil to its own return so that 160°F water enters the unit, and 140°F water leaves the unit and returns to the boiler to be heated up only 20°F to its normal 160°F supply temperature. The chilled water similarly flows in at 45°F and out at 55°F back to the chiller, where it needs to be cooled down only 10°F.

Fan coil units can be used on perimeters handily because they are designed to sit below windows and direct a stream of warm air upward to counter the drafts falling downward. In cooling season, adjustable louvers are often used to direct the air pattern toward the center of the room. Small, "through-wall" openings can be

FIGURE 7–14   The perimeter of this office building will be conditioned by a four-pipe fan coil system. The left photograph shows one of the floors in final stages of construction with the fan coil units in place. The right photograph shows the water supply lines with two-zone valves in place. One valve will control the hot water supply, the other the chilled water supply.

made and fresh air can be drawn in to meet the ventilation air requirement. Although in hotel and motel applications the controls may be mounted on a console in the unit, in office applications, where several such units are in one conditioned space, a remote wall-mounted thermostat may be used.

In some modern applications using floor to ceiling glass walls or sliding doors, a horizontal fan coil may be mounted in a dropped ceiling with a small amount of supply and return ductwork. The perimeter effect is not as great with this arrangement although linear diffusers in the ceiling over the windows directing the air downward could do a reasonable job.

7.4 OTHER PIPED SYSTEMS

The unit ventilator is a fan coil unit that was originally designed for schoolroom application. It provides heat in winter, using hot water or steam, and provides a degree of cooling by using outside air. Since school is usually in session in all but the hottest months of the year, the unit ventilator does a fairly decent job of providing heating and cooling as required.

Figure 7–15 shows the controls of the unit ventilator. In heating season the damper motor sets the outdoor air dampers to provide minimum air for ventilation. The thermostat energizes the valve, which could be either two-position or proportional, to control heating. As the room temperature rises, the valve shuts off. If the weather is mild and the temperature continues to rise, the signal from the thermostat energizes the damper motor to mix outside and

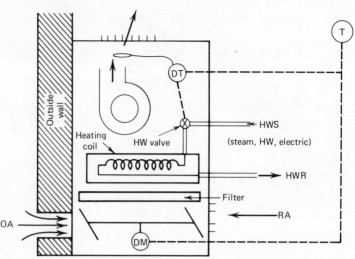

FIGURE 7–15   The unit ventilator provides fresh air for ventilation and for low cost cooling. It also has a heating function built in, which can be hot water, steam, or electric. It is commonly found in applications such as schools, hospitals, and other public buildings, where some cooling is desired but the full cost of mechanical refrigeration is not desired.

return air. At maximum cooling demand the outdoor damper is full open and the return air damper is full closed.

Notice the discharge air thermostat. It will sense the air temperature entering the room and if it drops too low, say below 55°F, it will open the heating valve to temper the air somewhat. In recent years the unit ventilator has been modified to reflect changing school-usage patterns and energy concerns. This equipment is now available with a refrigeration system built in to provide comfort during the summer season. Also, configurations employing electric heat and heat pump systems are being marketed.

A unit heater is a coil with a fan blowing air across it designed for space heating in factory, warehouse, or other areas where exceptional comfort coupled with esthetics is not important. Figure 7–16 shows such a device. The thermostat may cycle the fan although in some applications the fan may run constantly and the water and steam valve will be cycled. Since close temperature tolerance is not a major requirement, heavy duty line voltage thermostats are commonly used. Note also the thermostat sensing the temperature of the supply water. If the water is not hot enough to provide useful heating, the fan will not run even though the thermostat might be calling for heat. Unit heaters of this type are also available in gas-fired or electric heating configurations and are widely used in noncritical heating applications.

A common piped system for perimeter heating employs fin tube radiation. Copper tubes with aluminum or steel fins are enclosed in a

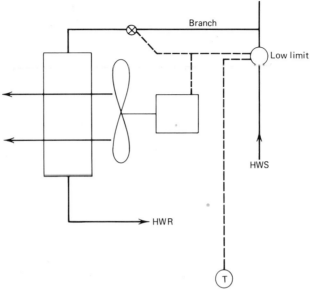

FIGURE 7–16   In this unit heater, the space thermostat senses a requirement for heat and sends a signal through the low limit thermostat to the water valve and the fan motor. If the low limit senses that hot water is available, the fan will start and the valve will open admitting hot water to the coil.

plain sheet-metal louvered cover. In some office applications rather decorative covers are available with extruded aluminum grille work, a selection of baked enamel finishes, or woodgrain vinyl cladding. Control of this system can be achieved with zone valves connected to wall-mounted thermostats, or sometimes with self-contained zone valves with thermostatic control built in, or again by a central panel responsive to the outdoor temperature. This panel regulates the water temperature as outdoor conditions change. At a certain temperature, say 50°F, the main circulator will come on. As the outdoor temperature drops, the supply temperature increases. If properly set and maintained, such a system can be effective. The building operating personnel must know how it works and how to adjust it. Many such systems are sources of continual discomfort because of over- or underheating.

7.5 COMBINATION SYSTEMS

Strictly speaking, most HVAC systems are combinations of air and water or ducted and piped systems although typically one or the other is dominant. In the case of the induction system we truly have a combination system. Figure 7–17 depicts an induction terminal. It is similar to a fan coil and is available in two-, three-, and four-pipe configurations. The terminal, however, does not have a fan. Air from a central air handler is ducted to the terminal and

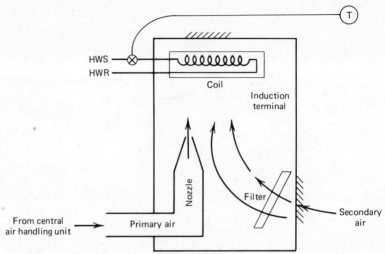

FIGURE 7-17   The coil in this figure is designed for heating only although three- and four-pipe heating/cooling configurations can be used. The action of the primary air passing through the nozzle draws or induces secondary air to flow into the unit from the room, hence the name "induction" system. Interior as well as perimeter areas can be served by such units.

passes through a nozzle in the unit. The Venturi effect of the nozzle draws room air through a grille at the lower front. The mixture of primary air from the main air handler and secondary air drawn into the unit passes over the coil, where additional heat is either added or removed. The major building load is handled by the central system as is the introduction of ventilation air. The fine tuning required at each terminal is accomplished by the coil. Since fresh air is being introduced to each room, a return air system must be included as part of the installation. The temperature of the primary air is usually controlled by a reset thermostat sensing the outside temperature. Although heating and cooling are theoretically available year round in the four-pipe version, in practice this is only true in spring and fall, with complete changeover being accomplished for the heating and cooling seasons.

In discussing the systems available to control a building's environment, we have seen a number of different approaches. Since a building has a core and a perimeter, it is common to have two systems in the building, often totally independent of each other. In addition to the two main systems, individual spaces may have unique requirements indicating other approaches. Entry halls may have a unit ventilator; stairwells may have a unit heater; and a computer room may have a small self-contained unit or units independent of any other equipment. The controls person must be able to identify the combinations of systems used and adjust the controls accordingly. It is not too farfetched to imagine a building with a perimeter system heating and a core system cooling, with the controls for each only inches apart. The building owner might as easily be throwing his energy dollars down a drain.

7.6 HEAT PUMPING   As mentioned previously, the core load is usually a cooling load, while the perimeter might require heating or cooling. The heat extracted from the core usually finds its way into the atmosphere either through a cooling tower or air-cooled condenser. If this heat could be transferred to the perimeter when heat is required, then an energy saving could be realized.

The growing use of incremental heat pumps makes this possible. Although heat is rejected to condenser water in the water-cooled core units, the temperature of this water is only about 95 to 100°F. Water at this temperature would not transfer heat too effectively if it were pumped to direct radiation equipment, but water at this temperature is ideal for a water-to-air heat pump. Figure 7–18 shows a control system employing water-cooled core units with water-to-air perimeter heat pumps.

In this scheme constant temperature water is circulated year round throughout the system. The typical water-to-air heat pump operates at water temperatures in the range of 60 to 90°F. The typical water-cooled cooling unit requires water at a temperature in the neighborhood of 85°F. Thermostat T has a set point of about 80°F. During the height of the cooling season, when both the core and perimeter units are calling for cooling, the water temperature in the loop will tend to rise. The thermostat senses this and turns on the cooling tower. The water-to-water heat exchanger then has cooling water running through it to keep the water temperature in the loop relatively low; that is, it directs the heat from the building out to the tower and from there into the surrounding air. In the height of the heating season, when all perimeter heat pumps are in the heating mode, the loop water will tend to drop in temperature. Thermostat T then will energize the hot water generator, which might be gas- or oil-fired or have electric heaters, to add heat in order to maintain the 80°F temperature required. If the core units are in the cooling mode during this time, the heat they remove from the core will be delivered to the perimeter and the hot water generator will require less energy.

During the in-between seasons the tower may be called upon to operate or perhaps the hot water generator may have to add heat to the water upon occasion, but more often than not the building will be in some sort of balance where the heat rejected by the core cooling units will match, more or less, the heat required by the core, in which case no additional energy need be added to the

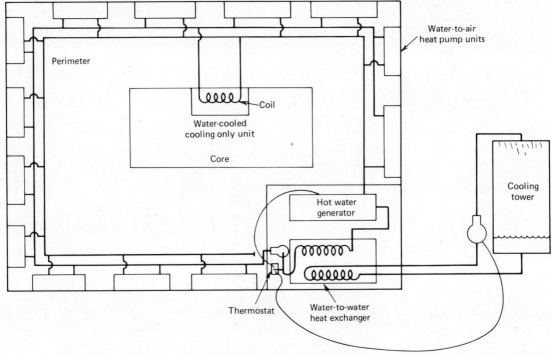

FIGURE 7–18   In this example of heat pumping, core heat is removed by the cooling unit and added to the water that is carried to the perimeter heat pumps. They in turn remove the heat from the water and add it to the air around the perimeter. This provides excellent economy during seasons when heating and cooling are needed simultaneously in the building.

building. This is the ultimate in heat pumping. An extra feature of such a system might be a storage tank that will accumulate heat during the day when most units are in the cooling mode, and release this heat at night when most units are in the heating mode. Obviously, the greater the flexibility of the system, the more elaborate will be the control system with increasing maintenance requirements.

7.7 CENTRAL PLANT

The last part of the HVAC system to be designed should be the central plant, that is, the equipment that will provide the heating and cooling. Not all systems require a central plant, for example, those using through-wall heating/cooling units; however, for the most part, some sort of central equipment room is usually required. The concept of district heating and cooling takes this to an extreme, where the central plant provides hot and cold water for a number of buildings in fairly close proximity to one another. Hospital and university complexes are good examples of such systems. Some experimentation is going on in which unrelated buildings are serviced by a central plant with each building paying for the Btu's of heating or cooling consumed. Steam has been available from central plants in some urban areas for a number of years.

The central plant is considered last in the design process because it must be designed to meet the requirements of the heating and cooling distribution system, which in turn has been designed to meet the functional requirements of the building. The central plant exists to service the building and its selection should reflect this. In retrofit work this may very well not be the manner in which the system is designed. In that type of construction a perfectly adequate existing central plant may dictate the manner in which the rest of the HVAC system is to be designed.

Figure 7–19 depicts a central plant capable of providing either hot or chilled water, but not simultaneously. Its major benefit is simplicity. Changeover from heating to cooling could be accomplished automatically, but in practice a manual system might be preferred. It is conceivable that an automatic system could change over several times per day as the temperature fluctuates during in-between seasons. This can be very expensive and, depending on how large a system we have, it may never provide the

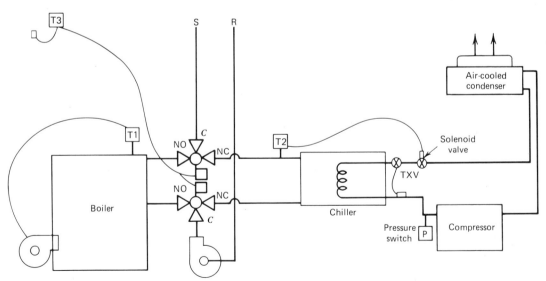

FIGURE 7–19   In the central plant, T1 maintains the boiler water temperature while T2 is designed to maintain chilled water temperature. Thermostat T3 is an outdoor control that operates the two-position three-way valves to change the system over from heating to cooling. Often manually operated valves are used for this purpose. Note that the chiller system shown is of the air-cooled type with pumpdown controls.

desired water temperature. More commonly, building operating personnel will arbitrarily switch over as weather changes even though there might be a few uncomfortable days in store for the occupants of the building.

The system shown in Figure 7–19 would be appropriate for single-zone, two-pipe fan coil systems. By splitting this system, as shown in Figure 7–20, it would be appropriate for multizone and four-pipe systems as well. The central plant is made up of two independent systems, one for cooling and one for heating. They may each run year round or be shut down as deemed appropriate by operating personnel. Diagrammatically, the various components of the central system are shown close together, but in fact they may be widely scattered throughout the building. A basement boiler location connected to a rooftop penthouse refrigeration room connected with fan rooms located on any number of floors is quite common.

In Figure 7–21 the hot water boiler has been replaced by a steam converter. Where steam is economically available, either from a utility or from another central plant, perhaps used in an industrial process, the use of a converter is quite popular. System water passes through the converter or heat exchanger, where it is heated by the steam. System capacity control is achieved by a steam control valve responding to a thermostat sensing the water temper-

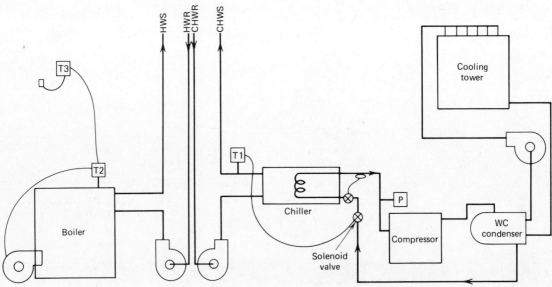

FIGURE 7–20   In this central plant, T3 is an outdoor thermostat that will reset T2 to maintain system water temperature at a level relative to the outdoor temperature. The water-cooled chiller system employs a pumpdown cycle controlled by T1, typically set at 45° F.

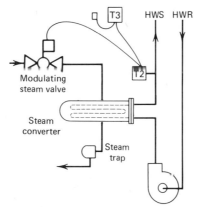

FIGURE 7–21   This steam converter is a heat exchanger designed to heat the water in the system by extracting heat from the steam. The steam trap shown insures that only condensate leaves the converter and that maximum heat is extracted from the steam delivered. The steam flow is controlled by the modulating valve in response to the system demand as measured by thermostat T2 and its master T3.

ature leaving the converter. The set point of this thermostat could be controlled by an outdoor reset controller.

In Figure 7–22 the central plant employs direct expansion refrigeration with a hot water generator (hot water boiler). The boiler supplies hot water to a perimeter radiation system at a temperature related to the outdoor temperature. It also supplies hot water to several multizone units. The hot deck temperature of these units is controlled by a three-way valve reacting to the zone thermostat sensing the greatest heating demand. The cold deck is basically uncontrolled except that as the system load decreases, unloaders cut out the cylinders of the compressor.

In designing the central plant, all elements of the system must be considered and accommodated.

7.8 ENERGY CONSIDERATIONS   Although it is desirable to have the HVAC system meet building requirements, it is even more desirable to make those requirements as energy-conservative as possible. Once that has been done, then the system can be designed to meet the requirements. Overdesigning a system, with the idea in mind that it can always be cut back once the system is in operation, is poor procedure. Equipment is designed for maximum efficiency at design conditions and usually operates at lower efficiency at off-design conditions. A control system designed for equipment of a particular capacity will generally not be able to control as effectively at off-design conditions either.

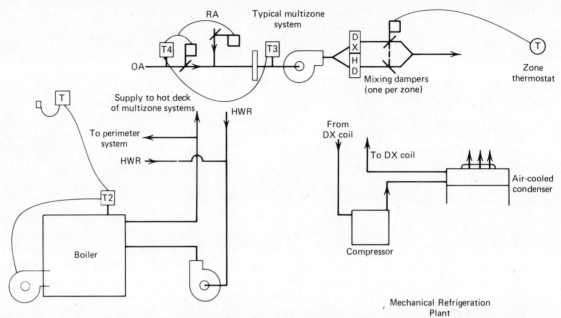

FIGURE 7–22   A shortcoming of the use of a DX coil is that the compressor and
condenser have to be fairly close to the coil to avoid losses in efficiency
due to long refrigerant piping runs. Many smaller systems do use DX coils
to good advantage in combination with hot water, steam, or direct-fired
heating equipment.

Energy conservation starts at the beginning of the functional
analysis of the building, whether a new construction project or a
retrofit job. A realistic indoor temperature in winter may be 70°F
while in summer 78°F may be adequate. Lighting levels based on
usage of the space rather than a blanket uniform level is an
effective means of reducing the cooling load. A lighting switching
system that permits selective lighting after normal business hours is
also important. Although it may sound trivial, it is not uncommon
to find buildings that require the lights on an entire floor to be on in
order for a porter to sweep a hallway.

Large auditoriums or spaces used infrequently or used seasonally
should have their own HVAC system, independent of the main
system. Offices used for longer than normal hours, and computer
facilities requiring 24-hour-a-day air conditioning should also have
independent systems. Large bulk storage areas should be kept as
cool as possible in winter and as warm as possible in summer.
Areas of physical activity should be kept 3 to 5°F cooler than areas
with sedentary activity. Ventilation is only required when the
building is occupied. Heating and/or cooling ventilation air in an
unoccupied office building is absurd. Similarly, exhaust systems
should be controlled based on usage.

In buildings requiring simultaneous heating and cooling, one should make sure that mechanical refrigeration is not cooling air that has just been warmed by the heating system. Overzealous energy conservation enthusiasts in the White House turned down thermostats during the heating season several years ago in an effort to demonstrate how energy could be saved. They found to their embarassment that the energy consumption went up, the system being a heating/cooling system that required more energy expenditure on cooling as the thermostat set point was dropped. Electric reheat in ducts carrying mechanically cooled air will also produce higher energy costs if thermostats are raised in summer. A central control system should have features that prevent such losses.

ASHRAE Standard 90–75, entitled Energy Conservation in New Buildings, analyzes building systems very carefully and gives guidelines for designing not only the building envelope but also the mechanical system. It is being increasingly referenced in municipal building codes as the source document for energy considerations.

DISCUSSION TOPICS

1. Describe how a single-zone system is controlled. Where might such a system be used to advantage?
2. Under what circumstances will a multizone ducted system be preferable to a single-zone system? Describe how it works.
3. What are some differences between a dual-duct terminal and a VAV terminal?
4. What are the pro's and con's of a four-pipe compared to a three-pipe fan coil system?
5. Where might a unit heater be used effectively?
6. How might the changeover from cooling to heating be accomplished? Why is a changeover control needed on VAV and fan coil systems?
7. How could analyzing the building usage result in HVAC operating changes that save energy?

Chapter Eight

Power Sources

In the previous chapters we have mentioned the sending of control signals from controllers to controlled devices without any further explanation of the nature of these signals. In this chapter we shall consider the three main types of control signals: electric, electronic, and pneumatic.

Electric control systems are commonly used in applications ranging from small residential through large commercial buildings. It is the most widely used type of system, and even in electronic or pneumatic systems it is not unusual to find electrically operated components.

Pneumatic systems tend to have a lower first cost only when the size of the system exceeds a certain point. Such systems are practically never found in small residential or commercial applications but are quite common in larger commercial and institutional buildings.

Electronic control systems are gaining in popularity in larger buildings because of their compatibility with data processing and computerized supervisory systems. Although both electrical and pneumatic control configurations can be adapted to computer use, the electronic system has a clear edge here.

8.1 ELECTRICAL SYSTEMS

Perhaps the most important part of understanding how electrical control systems work is understanding the electrical wiring diagram. By knowing what an electrical component is supposed to do, as well as in what sequence it is supposed to do it, one will be well on the way to mastering electrical controls.

The smallest part of a complex control layout that is of interest to us is the *circuit*. A circuit can be considered to be composed of four elements: (1) a power supply, (2) one or more switches, (3) one load, and (4) wires or conductors connecting everything together.

8.1.1 Power Supply

The source of electricity is measured in volts, frequency, and phase. In the United States the typical power supply has a frequency of 60 cycles per second or 60 *hertz* (Hz). This pertains to ac (alternating current) power supplies. In most electrical control systems we deal with ac power. We shall look at dc power when we get to electronic systems. Certain foreign countries use 50-hertz power, and there are electrical components available in this country made for 50-hertz systems. They are not compatible with our electrical control systems without modification. A typical 50-hertz electric motor will probably overheat and burn out or at least have a shortened life, as well as not perform as desired if an attempt is made to use it in a 60-hertz system. One should be aware of this limitation and make sure that the frequency rating of the components to be used is known before applying them to a particular system. Other than recognize that 50-hertz power exists, we need not consider it further because an understanding or lack of understanding of it will not affect our ability to comprehend electrical wiring diagrams.

The phase of the power supply is important in the same way as the frequency. We have single-phase and three-phase power available in HVAC systems. Three-phase power is used to directly drive motors and certain designs of electric heaters. *All control circuits use single-phase power.* Some small motors and electric heaters are also designed for single-phase power supply. For example, single-phase, air conditioning, hermetic compressors are available up to 5 tons in capacity; beyond that, with a few exceptions, only three-phase is available. Open motors generally only go up to about 1 or 2 horsepower as single-phase motors; beyond that they are available in three-phase only. Three-phase motors draw less current than single-phase motors in a given size although the power required is generally about the same. It becomes uneconomical to wire large single-phase motors with very heavy wire, when the same-size motor in a three-phase configuration will draw less current per leg and therefore require lighter, less expensive wiring.

The most important characteristic of the power supply from the viewpoint of the controls person is the voltage (sometimes called the "potential"). In electricity theory we saw that power consumed by an electric device is the product of voltage, current draw, and power factor ($P = V \times A \times$ P.F.). For a given-size electric load the current draw or amperes will drop as the voltage is increased. Heavy loads, that is, large horsepower motors and heaters, are powered by high voltage power supplies such as 460 or 220 volts. Light loads, such as the components used in control circuits, use lower voltage power supplies such as 115 or 24 volts. A way of thinking of voltage that may be useful in reading wiring diagrams is as a pressure forcing electrons through a wire. The higher the pressure (voltage), the greater the work that can be done.

Three-phase power supplies are represented in diagrams as shown in Figure 8–1. Note that there are three wires that supply power. The voltage will depend upon the generating equipment of

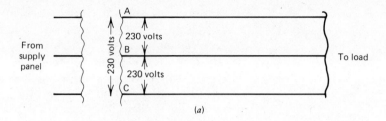

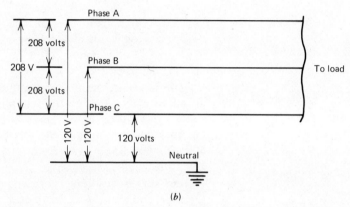

FIGURE 8-1   (a) A three-wire three-phase power supply sometimes referred to as a delta system; and (b) a three-phase four-wire power supply, often called a WYE system. Note that 120 volts can be obtained between any of the "hot" legs and the neutral wire.

the power company. In a three-wire three-phase power supply a voltmeter across each pair of "legs," or leads, will measure the same voltage; in fact, it must measure the same voltage within ±3% or the power company should be called to check it out. This voltage may be 240 or 220 or 208. In a three-phase, four-wire system, the voltage across each pair of hot legs will be 240 or 220 or 208, but the voltmeter between any of the hot legs and the fourth wire, or neutral, will read 110 to 125 volts.

The power supply used in a single-phase control circuit can be obtained in three ways. We may connect wires to two of the three hot legs of our three-phase power supply, which will give us about 230 volts, single-phase. We can connect one wire to a hot leg and another to the neutral leg giving us 115 volts single-phase. Or we can connect either the 230 volts or 115 volts, single-phase, obtained above, to the primary side of a stepdown transformer to get 24 volts. Figure 8-2 shows each of these techniques.

Note that the power supply is shown as two vertical lines in Figure 8-3. These are the uprights of what will shortly become a ladder-type schematic wiring diagram.

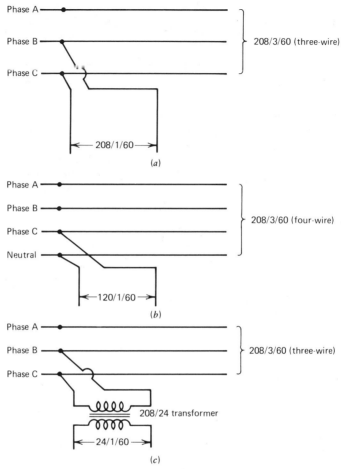

FIGURE 8–2 Common techniques for obtaining single-phase power for use in control circuits.

8.1.2 Conductors

The conductors or wires carry the electrons of our power supply much the way a pipe carries water. If a pipe is too small, the pressure drop through it will be excessive. In electricity, if the wire is too small, the voltage drop will be excessive. In order for an electrical device to work properly, the voltage supplied to it must be within ±10% of that for which it was designed. In circuits powering motors an excessive voltage drop will result in high current draw that could burn the insulation on the wires and might, in the extreme, lead to a fire. Oversized wires are not usually a problem except that they cost too much—who wants to spend money needlessly? In a circuit with an electric heater a high voltage drop will not result in high current, but rather the heater will just not get as hot as it should.

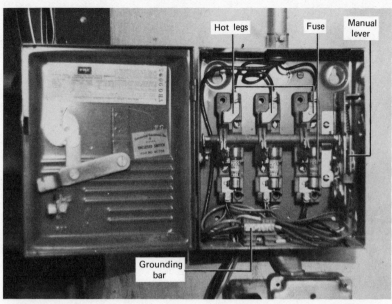

FIGURE 8–3   A disconnect box for a three-phase four-wire power supply. The three hot legs are fused with cartridge-type fuses. The neutral leg is wired directly to a grounding bar at the bottom of the box. Disconnects such as this should be used for any electric equipment. The limitations and applications of this box are written on the instructions attached to the inside of the cover.

Most of the conductors encountered in control systems are made of copper. Aluminum has been used in larger wires to power loads and in residential lighting, but most control wiring is still copper. Surrounding the conductor itself is a coating or insulation intended to prevent the loss of electrons. An electric circuit is complete when electrons start at a source of power and travel through the conductors, switches, and load to the source or ground. An uninsulated conductor touching a grounded part could provide a path for the electrons to follow other than that desired. Such a condition is called a *short circuit* and is minimized by the presence of sound insulation. The insulation may be made from any of a number of materials depending on how the wire is going to be used. Figure 8–4 lists some designations of wire insulation and typical applications.

One other characteristic of the conductor is its current-carrying capacity or ampere rating. The larger the size of the wire, the greater is its ability to carry electrons without excessive voltage drop. Figure 8–5 gives the size designations of wire for particular current ratings and line lengths. Note that if the length of the run from the power source to the point of use of electricity is great, a larger-size wire may be required.

A final word about conductors is that they must be protected. The National Electrical Code requires that some means be provided to protect the wires from overheating, burning, and causing a fire or other hazardous condition. The two most common means encountered are the fuse and the circuit breaker. Figure 8–5 also indicates the maximum circuit protection that must be provided for the conductor. The fuse is a one-time device, that is, it "blows," breaking the circuit, and must be replaced before power can be restored. The circuit breaker opens a circuit in response to excessive current, but it can be manually reset, restoring power. Both are widely used in control circuitry.

8.1.3 Loads

A load is a device that consumes electrical energy and in the process does work. Technically speaking, the electrical energy is converted to another form of energy, but for the most part we can consider the load as consuming energy.

There are two broad categories of loads in which we are interested, namely, inductive and resistive. Electric heaters and lights are examples of resistive loads. If we were to measure the current going to such a load at the moment the load is energized, we would see the current rise quickly to a maximum value and stay there. Taking the same measurement with an inductive load, we would find that the current would peak quickly and then drop back to a value one-third to one-fifth of the peak value. This peak is called inrush current, or in the case of motors, locked rotor current. Electric motors and relay coils are examples of inductive loads.

Since the purpose of an electric circuit is to do a job by energizing a load, the load must be carefully selected to accomplish this purpose. The voltage, frequency, and phase of the load must be the same as the available power supply. The conductors connecting the power supply to the load must be large enough to carry the required current. On any load you should be able to find a nameplate that indicates the characteristics of the required power supply as well as the current (amperes) that the load will draw. In the case of resistive loads you might find that the power consumption in watts rather than current draw is given.

8.1.4 Switches

Perhaps the most important elements in the circuit from the point of view of controls people are the switches. The switches determine the timing and sequence of operation of the loads. It is rare that a circuit has no switches at all, since this implies that the load is always energized. Circuits with a half-dozen or more switches are not at all uncommon.

A switch is a device that starts or stops the flow of electrons to the load, much the same as a valve in a water line stops or starts

Trade Name	Type Letter	Max. Operating Temp.	Application Provisions
Heat-Resistant Rubber-Covered Fixture Wire	*RFH-1	75°C 167°F	Fixture wiring. Limited to 300 V.
Solid or 7-Strand	*RFH-2	75°C 167°F	Fixture wiring, and as permitted in Section 310-8.
Heat-Resistant Rubber-Covered Fixture Wire	*FFH-1	75°C 167°F	Fixture wiring. Limited to 300 V.
Flexible Stranding	*FFH-2	75°C 167°F	Fixture wiring, and as permitted in Section 310-8.
Thermoplastic-Covered Fixture Wire—Solid or Stranded	*TF	60°C 140°F	Fixture wiring, and as permitted in Section 310-8.
Thermoplastic-Covered Fixture Wire—Flexible Stranding	*TFF	60°C 140°F	Fixture wiring.
Cotton-Covered, Heat-Resistant, Fixture Wire	*CF	90°C 194°F	Fixture wiring. Limited to 300 V.
Asbestos-Covered Heat-Resistant, Fixture Wire	*AF	150°C 302°F	Fixture wiring. Limited to 300 V. and Indoor Dry Location.
Silicone Rubber Insulated Fixture Wire	*SF-1	200°C 392°F	Fixture wiring. Limited to 300 V.
Solid or 7 Strand	*SF-2	200°C 392°F	Fixture wiring and as permitted in Section 310-8.
Silicone Rubber Insulated Fixture Wire	*SFF-1	150°C 302°F	Fixture wiring. Limited to 300 V.
Flexible Stranding	*SFF-2	150°C 302°F	Fixture wiring and as permitted in Section 310-8.
Code Rubber	R	60°C 140°F	Dry locations.
Heat-Resistant Rubber	RH	75°C 167°F	Dry locations.

*Fixture wires are not intended for installation as branch circuit conductors nor for the connection of portable or stationary appliances.

FIGURE 8–4   This table, extracted from the National Electrical Code, identifies various conductor insulation materials. It lists the maximum operating temperatures and restrictions on applications.

Trade Name	Type Letter	Max. Operating Temp.	Application Provisions
Heat Resistant Rubber	RHH	90°C 194°F	Dry locations.
Moisture-Resistant Rubber	RW	60°C 140°F	Dry and wet locations. For over 2000 volts, insulation shall be ozone-resistant.
Moisture and Heat Resistant Rubber	RH-RW	60°C 140°F 75°C 167°F	Dry and wet locations. For over 2000 volts, insulation shall be ozone-resistant. Dry locations. For over 2000 volts, insulation shall be ozone-resistant.
Moisture and Heat Resistant Rubber	RHW	75°C 167°F	Dry and wet locations. For over 2000 volts, insulation shall be ozone-resistant.
Latex Rubber	RU	60°C 140°F	Dry locations.
Heat Resistant Latex Rubber	RUH	75°C	Dry locations.
Moisture Resistant Latex Rubber	RUW	60°C 140°F	Dry and wet locations.
Thermoplastic	T	60°C 140°F	Dry locations.
Moisture-Resistant Thermoplastic	TW	60°C 140°F	Dry and wet locations.
Moisture and Heat-Resistant Thermoplastic	THW	75°C 167°F	Dry and wet locations.
Moisture and Heat-Resistant Thermoplastic	THWN	75°C 167°F	Dry and wet locations.
Thermoplastic and Asbestos	TA	90°C 194°F	Switchboard wiring only.
Thermoplastic and Fibrous Outer Braid	TBS	90°C 194°F	Switchboard wiring only.

Allowable Current-Carrying Capacities
of Insulated Copper Conductors in Amperes

Not More than Three Conductors in Raceway or Cable or
Direct Burial (Based on Room Temperature of 30° C. 86° F.)

Size AWG MCM	Rubber Type R Type RW Type RU Type RUW (14-2) Type RH-RW See Note 9 Thermoplastic Type T Type TW	Rubber Type RH RUH (14-2) Type RH-RW See Note 9 Type RHW Thermoplastic Type THW THWN	Paper Thermoplastic Asbestos Type TA Thermoplastic Type TBS Silicone Type SA Var-Cam Type V Asbestos Var-Cam Type AVB MI Cable RHH†	Asbestos Var-Cam Type AVA Type AVL	Impregnated Asbestos Type AI (14-8) Type AIA	Asbestos Type A (14-8) Type AA
14	15	15	25	30	30	30
12	20	20	30	35	40	40
10	30	30	40	45	50	55
8	40	45	50	60	65	70
6	55	65	70	80	85	95
4	70	85	90	105	115	120
3	80	100	105	120	130	145
2	95	115	120	135	145	165
1	110	130	140	160	170	190
0	125	150	155	190	200	225
00	145	175	185	215	230	250
000	165	200	210	245	265	285
0000	195	230	235	275	310	340
250	215	255	270	315	335	....
300	240	285	300	345	380	....
350	260	310	325	390	420	....
400	280	335	360	420	450	....
500	320	380	405	470	500	....
600	355	420	455	525	545	....
700	385	460	490	560	600	....
750	400	475	500	580	620	....
800	410	490	515	600	640	....
900	435	520	555			
1000	455	545	585	680	730	....
1250	495	590	645	....	....	....
1500	520	625	700	785	....	....
1750	545	650	735		....	....
2000	560	665	775	840	....	....

CORRECTION FACTORS, ROOM TEMPS. OVER 30° C. 86° F.

C. F.						
40 104	.82	.88	.90	.94	.95	....
45 113	.71	.82	.85	.90	.92	....
50 122	.58	.75	.80	.87	.89	....
55 131	.41	.67	.74	.83	.86	....
60 140	....	.58	.67	.79	.83	.91
70 158	....	.35	.52	.71	.76	.87
75 167	....	....	.43	.66	.72	.86
80 176	....	....	.30	.61	.69	.84
90 194	....	....	....	.50	.61	.80
100 212	....	....	....	....	.51	.77
120 248	....	....	....	....	....	.69
140 284	....	....	....	....	....	.59

†The current-carrying capacities for Type RHH conductors for sizes AWG 14, 12 and 10 shall be the same as designated for Type RH conductors in this Table.

FIGURE 8–5   This table, extracted from the National Electrical Code, lists wire sizes (AWG designation) and current-carrying capacity in amperes. A restriction to this table is that no more than three conductors are in the conduit or raceway. This is the table most frequently used in selecting HVAC application wires.

water flow. Taking the similarity between water and electrons one step farther, we might consider that a valve in an intermediate position (neither full open nor full closed) can modulate the flow of water. There are electrical equivalents of such a valve, called variable resistors or potentiometers, that are used to only partially block the flow of electrons to the load. In a sense we can consider these devices switches although commonly switches are thought of as only stopping or starting the full flow of electricity.

A controls person, in addition to being interested in what happens when a switch opens or closes, is perhaps even more interested in *why* the switch opens and *when* it opens. A switch that opens as a result of a temperature change is called a thermostat or temperature controller. A switch that opens as a result of a pressure change is a pressurestat or pressure switch or pressure control. A switch that opens in response to a change in humidity is called a humidistat. A switch that opens in response to an electromagnetic force created by electrons flowing through a coil of wire is called a relay. Incidentally, in the HVAC field generally, a relay is considered to be a fairly small switch, that is, capable of controlling a load with a rating of perhaps 10 to 20 amperes. Beyond that size such a switch is usually called a contactor.

We can continue to list the types of switches. One that responds to a change in current going to a motor is called an overload switch. If the switch is opened by a cam connected to the shaft of a clock motor, it is called a timer switch. If it opens several seconds or minutes after called upon to do so, it is called a time delay switch. The switches are almost limitless in their application, and it is this fact that really makes the controls field fascinating and challenging.

In applying switches of whatever sort to a particular application, the rating of the switch must be kept in mind. The parts of the switch subject to the most wear and tear are the contacts, the points that come together to close the circuit or open to break the circuit. The ability of the contacts to do the job is reflected in the voltage and current rating. Heavy, substantial contacts can carry more current and open cleanly against a higher voltage than can light delicate contacts. Unfortunately, heavy, substantial contacts are more costly and less responsive to, say, temperature change assuming that the switch under discussion is a thermostat; therefore, all contacts cannot be made heavy and substantial. The more responsive a switch is, the lighter and more delicate it must be. There are ways around this requirement as we shall see shortly. The main point is to make sure the voltage and current rating of the switch is equal to or greater than the voltage and current rating of the load it is to control. Incidentally, some switches are rated for both inductive and resistive loads. This information is on the

Highlights—Fuses

The primary job of a fuse according to the National Electrical Code (NEC) is to protect the conductors in an electrical circuit. The requirements of the NEC are reflected by the types of fuse likely to be encountered in the HVAC field today. Several fuses are shown in Figure 1.

Two broad categories of fuse are the ordinary or "quick blow" fuse and the time delay fuse. Since ac motors are characterized by a high inrush current at start-up, a quick-blow fuse used in motor circuits may cause nuisance trips. Such a fuse, however, may be quite adequate for electric heat and lighting circuits. The time delay fuse will respond rapidly to short circuits but will tolerate the high inrush currents of motors at start-up without blowing.

Another means of categorizing fuses is by their appearance. Plug fuses are not available for use in systems with voltages greater than 150 volts or with a current draw in excess of 30 amperes. They are available, though, in both quick-blow and time delay configurations. The standard plug fuse has a screw connector referred to as an "Edison base" for use in the typical fuse box. The tamper-proof plug fuse shown here has been designed with an undersized base to prevent a penny from being placed behind a blown fuse and bridging the gap and thereby eliminating the circuit protection. The adapter shown permits use of the tamper-proof fuse in standard Edison base fuse boxes.

Cartridge-type fuses are available in three distinct configurations. The glass tube type is commonly found in control circuits up to a maximum of 250 volts at very low current ratings. The cartridge-type *without* blades can be used up to 250 volts and 60 amperes; the cartridge type *with* blades can be used up to 600 volts and 600 amperes.

Fuses are selected based on voltage and current-carrying capacity of the conductors as set forth in the NEC. See Figure 8–5 of the text for an example of a table found in the NEC. Smaller time delay fuses may be used to provide protection to the motor in the circuit if desired. For example, a number 14 AWG conductor is wired to a motor with a rating of 9 amperes. The wire has a maximum rating of 15 amperes; however, a 10-ampere fuse was selected because it will protect the conductors and also provide a measure of overcurrent protection for the motor.

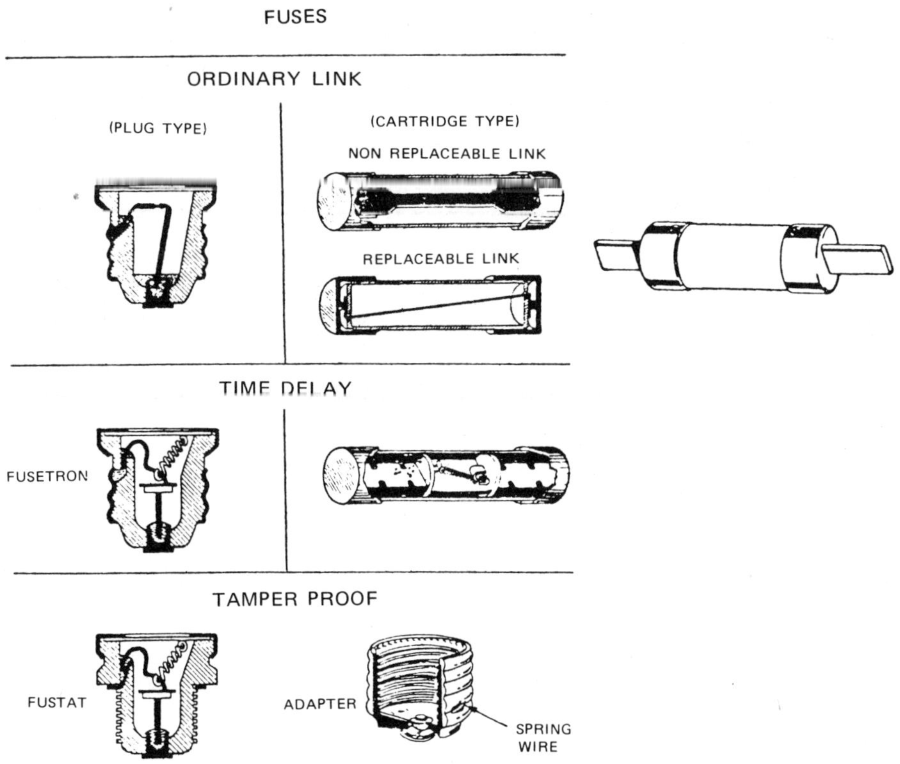

FUSES

ORDINARY LINK

(PLUG TYPE)

(CARTRIDGE TYPE)

NON REPLACEABLE LINK

REPLACEABLE LINK

TIME DELAY

FUSETRON

TAMPER PROOF

FUSTAT

ADAPTER

SPRING
WIRE

FIGURE 1

FIGURE 8–6   Two time clocks. At the right is a 24-hour clock used to energize and deenergize equipment at a fixed time during the day. At the left is a 7-day clock commonly used in HVAC systems to turn on the equipment in the morning, shut it off at night, and insure that it stays off over the weekend. As you can see from the dial face, any program over a 7-day period can be selected. (Courtesy of Johnson Controls Inc.)

switch rating plate and merely means that one should consider the type of load being controlled as one attempts to apply the switch to the particular job.

8.1.5 The Ladder Schematic Diagram

In order for a building designer to communicate with the construction crew in the field, a set of drawings and specifications is created. To enable the controls designer to communicate with the electrician and controls service person in the field a wiring diagram and a sequence of operations are created. The ladder schematic is a form of wiring diagram intended to show the *electrical* relationship of the power supplies, conductors, loads, and switches that make up the control system. A pictorial diagram attempts to show the *physical* relationship of the various components, and it is quite

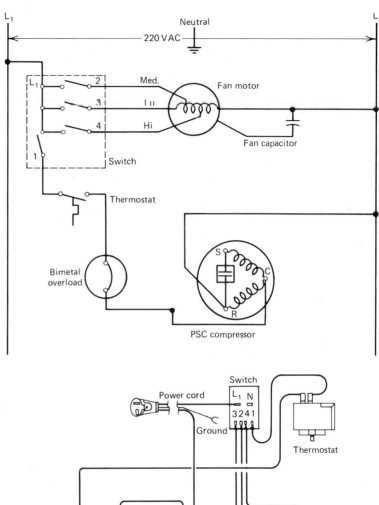

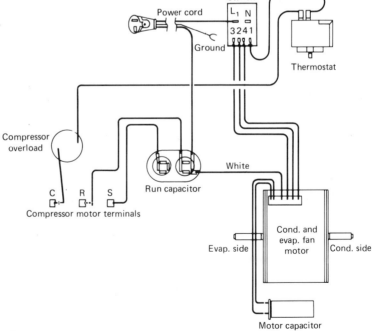

FIGURE 8-7 Two wiring diagrams for a room air conditioner. The ladder diagram is preferred for electrical troubleshooting; the pictorial diagram is useful in locating components and identifying wires and where they connect.

useful in locating the parts in a building or in a complex piece of HVAC equipment. It is generally not too useful in troubleshooting or in determining system operation. Both diagrams are very helpful; if it is necessary to make a choice, the ladder schematic is more important.

Figure 8–8 shows some symbols commonly used in schematics. An industry standard for wiring diagrams does exist, but not many companies follow it. Fortunately, there is a sufficient overlap between varying techniques, and enough marginal notes, to help one who is conversant with a particular style of diagram make out a different-style diagram fairly readily. The symbols shown here are common enough to make committing them to memory a worthwhile exercise.

The ladder schematic in Figure 8–9 may seem a bit horrendous to someone with no experience at all in reading such diagrams. Actually, it is nothing more than a bunch of circuits of the sort previously described gathered together in one diagram. In attempting to make sense out of the schematic, one must have more than just the ability to read schematics. It is essential to know how the equipment or system under study is intended to operate and what each individual component is supposed to do. It is not enough to identify a pressure control, but rather one must identify it as a pressure control designed to open on a decrease in pressure caused by a liquid line solenoid valve shutting off in response to the space thermostat. *One must know the system.* This knowledge only comes from studying *all* the documents such as wiring diagrams, instruction manuals, the sequence of operations, specifications, and any other information from whatever source pertaining to the system. The ability to read wiring diagrams well comes after a good bit of experience in the field.

In Figure 8–9 note some of the major characteristics. The two vertical lines identify the control circuit power supply; the horizontal lines are connected to the loads. In some cases more than one line leads to the same load, and in other cases one line leads to more than one load. Each load is identified by a symbol, which is further defined in the legend at the right of the diagram. Once we have identified each load and understand its purpose, we are ready to identify the switches in each circuit and how they relate to the load. This orientation is vital in any study of diagrams. We always start with the overall picture and gradually narrow our view to individual loads and switches as we increase our understanding of the purpose of the entire system.

8.1.6 The Circuit

We can isolate one circuit in the diagram shown in Figure 8–10 and look at it in detail. Notice that the power supply originates at L1 and terminates at L2. It is important to remember that current

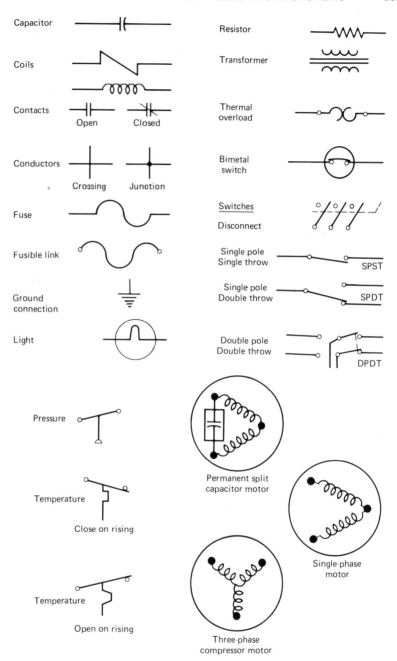

FIGURE 8–8   Symbols commonly used throughout the HVAC industry. Although variations exist, they are sufficiently similar so that confusion should not be great.

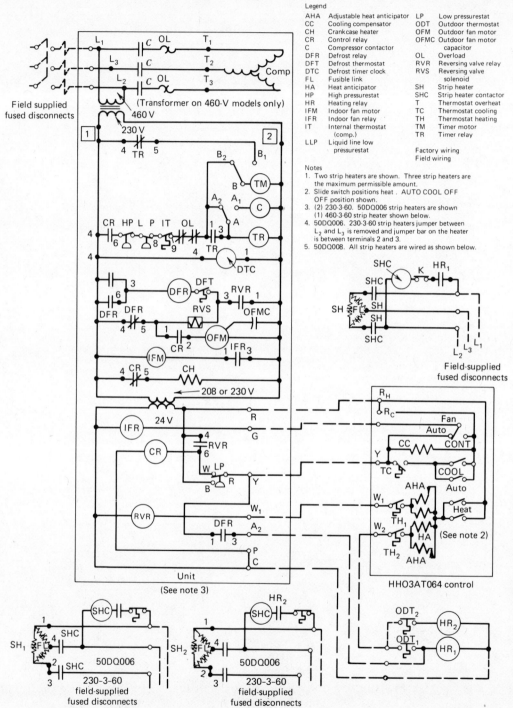

FIGURE 8–9   A schematic wiring diagram typical of those encountered in the field. Note that the equipment represented by this drawing has 460/3/60, 230/1/60, and 24/1/60 components. By carefully studying this diagram, you can begin to break it down into loads and switches that make up each circuit. Once you have done this, you are on your way to troubleshooting.

flows from one leg to the other leg. Although in ac power the electrons flow back and forth 60 times each second at the speed of light, we need not concern ourselves with that seemingly magical fact. Let us just assume that electron flow is from left to right and that if we start at L1 we *must* end up at L2.

The load in this circuit is the indoor fan motor. It has a rating of 230 volts. In order for the motor to run, the circuit from L1 to L2 must be complete with no breaks. All the switches, in this case two of them, must be closed. Note that the symbol for the normally closed switch is OL. The legend identifies this as an overload and our knowledge of motors should tell us that this switch is probably sensitive to excess current and is intended to open, deenergizing the motor if the current becomes excessive. The expression "normally" closed usually means that when the system is deenergized, with nothing running, this switch will be in the closed position; that is, its contacts will be closed.

The other switch in the circuit is labeled R1. The legend identifies this as an indoor fan motor *relay*. The term *relay* should bring to mind an electrically operated switch with contacts and a coil. The contacts in this case are normally open and labeled R1, and the coil is a load not shown in this circuit. Although physically the coil

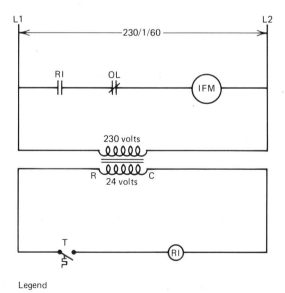

Legend

IFM = Indoor fan motor
RT = Fan relay
OL = Overload
T = Room thermostat

FIGURE 8–10   Part of a more complex wiring diagram for an air conditioning system. There can be any number of line voltage loads above the transformer and any number of low voltage loads below the transformer. In this case only the fan motor and fan relay coil are shown.

and contacts are in one piece of hardware, electrically they are in two different circuits. In order to get the indoor fan motor to run, we must complete the circuit from L1 to L2 through the two switches and the load. However, in order to close one of the switches, R1, we must complete a second circuit from *R* to *C* through a load labeled R1, which is the *coil* of the indoor fan motor relay.

In this second circuit we note that the power supply is no longer 230 volts across L1 and L2 but rather 24 volts across *R* and *C* of the transformer. This is called a *low voltage* circuit as opposed to the previous circuit we examined, which is called a *line voltage* circuit. Many electrical control systems use 24-volt low voltage circuits because the control elements can be light duty, responsive to small changes in control variables and relatively inexpensive to install. In the circuit from *R* to *C* through the R1 coil there is only one switch that can be identified by the symbol as a thermostat; from the legend it is identified as the room thermostat. As the temperature rises, the thermostatic switch closes completing the circuit through the R1 coil. This causes the R1 switch in the line voltage circuit to close completing the circuit to the indoor fan motor causing it to run.

The rest of the diagram can be similarly examined. Note that there is no magic involved here. An ability to read and interpret symbols, along with a knowledge of the individual components and how they operate and the sequence of operation are all that is required. This is not to suggest that this procedure is easy but rather that it can be done by anyone with persistence and a desire to be one of the most valuable people on the construction team.

8.2 PNEUMATIC SYSTEMS

The object of any control system is to sense the condition of a variable, analyze this condition, and cause a change to occur in the HVAC system that will cause the variable to assume a desired value. The sensor, controller, and controlled device that will do these things can be electrical, as we have just seen, or compressed air can be used as the source of energy as in a pneumatic system. The sensor can sense a change in a variable and send a pneumatic (compressed air) signal to the controller, which in turn analyzes the signal and sends still another signal, also air under pressure, to the controlled device. This device receives the signal and converts the pressure into a force that drives a damper, valve, or other mechanism.

8.2.1 Compressed Air

The air supply for the typical pneumatic control system uses a pressure of about 100 psig. This is produced by an air compressor at a central location and distributed by means of tubing, usually

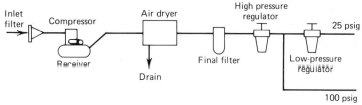

FIGURE 8-11   Components commonly found in the mechanical equipment room of a building using pneumatic controls. Ambient air is drawn into the compressor, compressed, dried, filtered, and passed on to the pressure regulators, where pressures are reduced to the level required by the particular system used in the building.

copper, to where it is needed. The compressor and some of its controls are shown in Figure 8-11. A requirement of this air is that it be clean and dry. Cleaning is accomplished in several stages. Often a prefilter will clean the air drawn into the compressor. A secondary filter, with a much greater filtering ability is located in the high pressure line to remove as many fine particles as possible. If such particles were to get into the small control orifices and passages, the controls would be useless. Very often individual control components will also have integral filters for final filtration.

Drying can be accomplished by one of several methods. Regenerative driers absorb moisture by means of adsorption. Periodically, they must be "regenerated," that is, heated in a way that will drive off the moisture and return it to the atmosphere, restoring the drier material to its original efficiency. Although some driers require manual removal and regeneration periodically, on larger systems the process is more often accomplished automatically. Mechanical refrigeration is a widely used drying technique. The compressed air is passed over a cold evaporator, and moisture is condensed out of it. Even in large buildings the amount of air used is rather small so that the refrigeration equipment is also rather small, often being less than 1 horsepower in size.

The pressure in the receiver or storage tank will rise when the compressor is on and fall when the compressor is off and air is being used. The pressure switch on the compressor cycles the compressor to keep the pressure between, say, 105 and 125 psig. The high pressure requirement in the system might be 100 psig. The high pressure regulator is designed to maintain a constant 100 psig even though the pressure in the receiver fluctuates between 105 and 125 psig. Since most of the control components in the pneumatic system are designed to operate at pressures of 15 or 18 or perhaps 25 psig, depending on the manufacturer of the equipment, a low pressure regulator is used to lower the 100 psig pressure to the required lower value.

The supply lines carrying the air to the controls are called mains, and the letter "M" is commonly used on control devices to show

FIGURE 8–12   A pneumatic system compressor having two belt-driven compressors mounted on a single receiver tank. Each compressor is capable of handling the building load alone. In the event of a problem with one of the compressors the second will be available. Both will be used alternately during normal operation to ensure even wear and proper functioning.

where this supply air should be connected. Increasingly, low-cost, easy to use, plastic tubing is being used in low pressure control applications. Lines that are not mains are generally called branches. Physically, they may be identical in appearance to the mains.

8.2.2 Pneumatic Actuators

The pneumatic actuator is the part of the system that does the work. As shown in Figure 8–15, it is nothing more than a piston in a cylinder. Air from the controller enters the space above the piston and causes the piston to move against a spring. The extent of the piston movement is governed by the strength of the spring and the amount of pressure coming from the controller.

FIGURE 8–13   A dryer of the refrigeration type intended to provide control quality air for the pneumatic system.

Typically, the piston will not move at all as the pressure increases from zero to about 3 psig; as the pressure increases above 3 psig, it will start to move. If the pressure stops at 5 psig, the piston will stop and hold; as the pressure again increases, it will continue to move and the piston rod will extend. At 15 psig the piston rod is fully extended. Proportional control with pneumatic systems is quite easy. All we need is a controller that can send a pressure signal that varies with the change in temperature. We have an actuator that is able to respond readily to such a proportional signal.

The force exerted by the actuator depends on the pressure signal and the area of the piston. As the force requirement increases, the diameter of the piston must also increase. This can lead to some fairly unwieldy designs so that an alternate to the piston-type actuator has been devised, namely the diaphragm actuator. Figure 8–16 shows a commonly used valve operator employing a large-diameter diaphragm. The force that can be created by a diaphragm

FIGURE 8–14   As the air moves from left to right, it is filtered and the pressure is reduced to required system operating levels.

measuring 15 inches in diameter having a pressure signal of 15 psig acting on it can readily be calculated by multiplying the diaphragm area by the pressure. In this case the force works out to be about 2650 pounds. Compare that to the 188 pounds of force that can be created by a piston-type actuator 4 inches in diameter.

There may be applications where a large force is required, but a diaphragm actuator may be inconvenient to use because of space limitations. The answer to such a problem is to increase the pressure rather than the piston diameter. Although the control system has a maximum pressure of only 15 psig, in some systems a high pressure main is installed with pressures up to 100 psig right along with the 15 psig control main. A device called a positive-positioner relay is designed to respond to the control signal in the

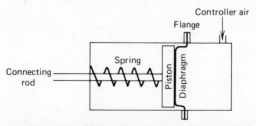

FIGURE 8–15   As the controller pressure varies, the force on the piston varies and the extension of the connecting rod varies proportionately. Note the diaphragm used to seal the gap between the piston and the cylinder walls. The spring is usually factory-set to respond to pressure variations over a prescribed range.

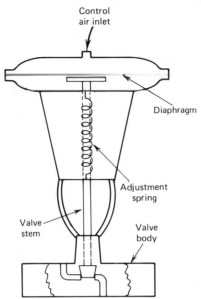

FIGURE 8-16   This valve operator uses a diaphragm in order to develop a force large enough to overcome the pressure exerted by the fluid in the valve. A pneumatic operator with a 15-psig operating pressure can often be used in fluid systems with operating pressures as high as 2000 psig.

range of 3 to 15 psig. As the signal increases, the relay in turn controls the 100 psig main supply and meters an air signal into the piston. If it receives a control signal of 3 psig, it maintains a zero pressure on the piston, and the piston rod is fully retracted. If it receives a 15 psig signal, it increases the pressure to the piston to 100 psig and the piston rod extends fully. An intermediate control signal of, say 9 psig, will produce a 50-psig pressure on the piston and the piston rod will extend halfway.

In the preceding discussion all the actuators extended as pressure was increased on the piston. Some actuators, and very commonly diaphragm-type actuators, are designed so that pressure is introduced in a manner that will cause the actuator arm to retract. This is a design feature; usually, an actuator is specifically designed for one type of operation or the other and is not convertible.

8.2.3 Pneumatic Controllers

The pneumatic controller is a device that receives air pressure from the main and modifies the pressure to produce a signal, the type of signal being dependent on its designed purpose. Typically, the signal will vary between a minimum value, usually about 3 psig, and a maximum value, usually about 15 psig.

If the signal goes to an indicating dial, we have an instrument. The dial may be calibrated in degrees Fahrenheit to convert the

(a)

FIGURE 8–17   Pneumatic actuator (a) is mounted on a water valve typical of those used in smaller fan coil or duct coil applications. Actuator (b) is used on large return or supply air dampers; actuator (c) is used for small damper applications such as unit ventilators. (Courtesy of Johnson Controls Inc.)

pressure signal to a temperature reading. The controller has a mechanism built into it that will convert a changing temperature into a changing pressure signal.

Figure 8–18 gives a schematic representation of a bleed-type controller responsive to temperature change. The supply air pressure of 15 psig is fed through the main to a restrictor. The restrictor

(b)

(c)

limits the amount of air that can enter the controller. This amount of air is so small that it is measured in cubic inches per second. The branch line is connected to the pneumatic actuator. If there is no means of escape for the air, all of it will flow to the actuator and the pressure in the branch line, also called the control signal, will reach 15 psig. The controller, however, has a nozzle that will bleed off some of the pressure. In fact, if the nozzle is completely open, all the air will escape resulting in a control signal of zero. As the nozzle becomes restricted by the bimetal element, less and less air

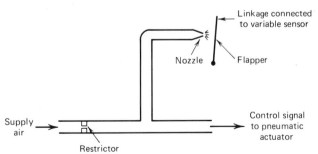

FIGURE 8–18   In this bleed-type pneumatic controller the linkage is connected to a sensor (pressure, temperature, humidity, etc.), which causes the flapper to allow more or less air to escape. The control signal sent to the actuator depends upon the clearance between the nozzle and the flapper, a clearance commonly measured in thousandths of an inch.

FIGURE 8–19   The pneumatic controller at the left is wall-mounted and contains a thermometer and adjustable set point. That at the right can be field-adjusted, but the cover must be removed first. Usually, a special tool is required to do this; hence the terminology "tamper-proof" is used to describe this device. (Courtesy of Johnson Controls Inc.)

can escape with the result that the pressure in the branch increases. If the bimetal element should completely cover the nozzle, the control signal would increase to the maximum 15 psig pressure. If this branch were connected to a dial indicator, as mentioned before, we would have a thermometer, or if it were connected to a pneumatic actuator we could open or close a damper or valve.

Since temperature, and in fact most variables, changes gradually, the change in the control signal is also gradual. This fact makes pneumatic control systems inherently proportional. Although electrical systems are inherently two-position on-off systems and require a fairly complicated control scheme to provide proportional control, the reverse is true of pneumatic systems. Two-position control can be achieved using a relay.

Before moving on to relays, it might be worthwhile to define several control terms commonly encountered in pneumatic systems. The expression *direct acting* is used to describe a controller that produces a control signal that changes directly as the variable measured. As an example, a pneumatic temperature controller that produces an increasing signal as the temperature rises is called a direct acting (abbreviated DA) thermostat. A *reverse acting* controller (RA) produces a *decreasing* pressure signal as the control variable increases; that is, in the thermostat mentioned, as the temperature increases, the pressure signal decreases and, as the temperature decreases, the pressure signal increases.

The *throttling range* of the controller refers to the change in the variable that causes a control signal change required to drive an actuator from the fully retracted to the fully extended position. If the valve, for example, is closed at a pressure of 3 psig and open at 15 psig when the controller senses temperatures of 75°F and 77°F, respectively, the controller is said to have a throttling range of 2°F. Said another way, a 2°F change of temperature will cause the valve

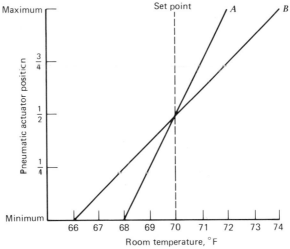

FIGURE 8-20   Throttling range. Control A has a throttling range of 4° F, control B of 8° F. Control A is more sensitive to temperature change as it causes the pneumatic actuator to move a greater part of its stroke as a result of a smaller temperature change. This is not always desirable because overshooting and undershooting may occur (sometimes called "hunting") and the system may never stabilize.

to go from full open to full closed. The larger the throttling range, the less sensitive a controller is to changes in the control variable. In the system just described a $\frac{1}{2}$°F change will cause the valve to move through $\frac{1}{4}$ of its travel while a controller with a 10°F throttling range would require a full $2\frac{1}{2}$°F change to cause the same $\frac{1}{4}$ travel of the valve.

A final word about controllers. They come in many shapes and sizes. The operating characteristics of the device are obtained from the manufacturer's specification sheets and operating instructions. Pneumatic controllers have the same flexibility as electrical controllers and are available in designs that respond to the pressure, temperature, enthalpy, humidity, and so forth. Different operating mechanisms may be used such as the bleed type described, nonbleed types, or others, but the applications are the same: They all sense a control variable and send a pneumatic signal to a control device.

8.2.4 Relays   A pneumatic relay is a device that receives a signal from a pneumatic controller or another relay and sends out a signal related in some way to the incoming signal. If there is any confusion in understanding pneumatic control systems, it usually arises in understanding the operation of relays. The internal workings of the relay are usually only of importance to the manufacturer of the relay. The person in the field is more concerned with the application and what changes occur to signals rather than how they occur.

One type of relay receives a direct acting signal from a controller and produces a reverse acting signal. As the pressure signal from the controller rises from 3 psig toward 15 psig, the signal from the relay drops from 15 psig toward 3 psig. In this way one controller can open one set of dampers and close another set at a different location. We shall consider some specific control configurations in a later chapter. At this point we are mainly concerned with introducing the fact that such devices exist.

Another relay changes a proportional pneumatic signal into a two-position signal. Another will redirect the flow of a pneumatic signal from one control path to another. A winter–summer changeover would be an application for such a device.

One type of relay will sense two incoming signals and send out a third signal proportional to the higher of the two; a similar device will send out a signal equal to the average of them both; and a third configuration will send out a signal proportional to their average.

Some pneumatic relays are referred to as interface controls. That is, they contain both a pneumatic and an electrical component. A pneumatic-electric relay is one that closes or opens an electric switch in response to a change in a pneumatic control signal. This operates in much the same way as a high or low pressure control on a refrigeration system. An electro-pneumatic relay is one in which an electrical signal causes a change in the pneumatic pres-

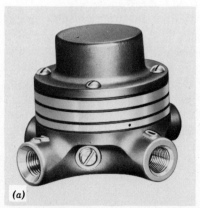

(a)

FIGURE 8–21   These relays receive pneumatic signals and act on them in a prescribed manner producing another pneumatic signal that goes to a controlled device or perhaps to still another relay. In the devices shown, the resulting signal can be proportional to a number of input signals, can be reversed, or can be the average of several input signals. (Courtesy of Johnson Controls Inc.)

(b)

(c)

sure. A simple solenoid valve controlling the airflow is an example of such a relay, often referred to as an EP relay.

The consensus among building operating engineers seems to be that pneumatic controls are best from the viewpoint of simplicity of operation, durability, reliability, and ease of troubleshooting. Of course, everything is simple once we know it, but it is generally true that pneumatic controls are the easiest to master. Since building operating engineers rarely make the decision about which control system will be used in a particular application, we find electrical and electronic systems widely used as well as pneumatics.

8.3 ELECTRONIC SYSTEMS

The power source in an electronic system is in the range of 0 to 20 volts dc. All the components in the electronic system—the sensor, the controller, and the controlled device—must be compatible; that is, they must be designed to operate in a particular system together. Rarely can components of different manufacturers be interchanged; in fact, rarely can components of different control configurations made by the same manufacturer be interchanged. Although the range of all such systems is between 0 to 20 volts dc, the voltage changes within the system that cause dampers to move and relays to function are more frequently measured in millivolts (thousandths of a volt).

FIGURE 8–22   The pneumatic-electric relay, or P-E switch, causes a set of electrical contacts to make or break in response to a change in pneumatic signal. (Courtesy of Johnson Controls Inc.)

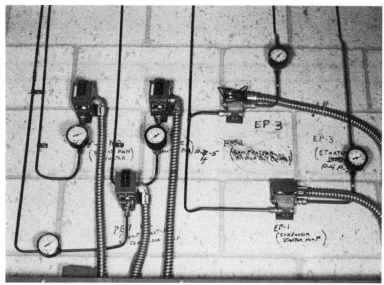

FIGURE 8–23   Building operating personnel have labeled, none too neatly, the pneumatic controls shown here. Three PE relays at the left are used to energize electric motors in response to pneumatic signals. Two EP relays at the right open in response to an electric signal and allow pneumatic control air to pass through to pneumatic controllers elsewhere in the building.

8.3.1 Temperature Sensing

It is the temperature sensor of the electronic system that dictates the makeup of the rest of the components in the system. Temperature sensing is accomplished by using a wire that has an electrical resistance that changes as the temperature changes. If the resistance of the sensor increases as the temperature increases, it is said to have a *positive* coefficient of resistance. If its resistance decreases as the temperature increases, it is said to have a *negative* coefficient of resistance.

The purpose of the rest of the equipment in our electronic system is to produce an electronic signal that will cause a controlled device to move in response to small changes in resistance that occur as the temperature changes.

8.3.2 Generating an Electronic Signal

The Wheatstone bridge is the heart of the electronic system. Its function is to compare resistances in the temperature sensors and to generate a signal that will provide corrective action in the HVAC system.

The bridge in its simplest form is shown in Figure 8–24. It consists of two circuits of two resistors each, powered by a dc source, in this case a 10-volt dc battery. Electricity will flow from X to Y through two parallel circuits, through R1 and R2, and through R3 and R4. Notice the voltmeter shown measuring the voltage across points A and B. The dc signal that we are looking for to cause corrective action is going to be generated across these two points.

Let us assume that R1, R3, and R4 are fixed resistors built into the main control panel of the electronic system and that each has a resistance of 10 ohms. We also assume that R2 is located in a remote location sensing the temperature of a conditioned space and that it has a positive coefficient of resistance. The temperature that it is sensing is such that its resistance is also 10 ohms. The bridge is said to be "balanced" and the voltmeter across A to B reads 0 volt.

Electrically, we can analyze the circuit to verify the zero reading. Using Ohm's law, we find that

$$I = \frac{E}{R}$$

The current through R1 and R2 is calculated thus:

$$I = \frac{10 \text{ volts}}{20 \text{ ohms}}$$
$$= \frac{1}{2} \text{ ampere}$$

The voltage drop across resistor R1 is then

$$E = IR$$
$$= \frac{1}{2} \times 10$$
$$= 5 \text{ volts}$$

Since we started with 10 volts and lost 5 volts potential in resistor R1, the remaining potential at point A is 5 volts. A similar calculation through R3 and R4 results in a similar voltage drop and the same 5-volt potential at point B. Since both point A and point B have the same potential, 5 volts dc, there is a zero potential difference between them, and hence a zero reading on a voltmeter, or a zero potential to do work.

Let us now assume that the temperature in the conditioned space increases. The resistance of R-2 also increases, say to 20 ohms. Again we analyze the bridge:

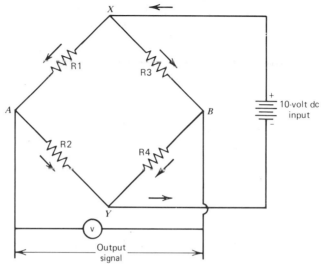

FIGURE 8–24   The Wheatstone bridge is the basic building block upon which electronic control systems are based. Rather complex electronic circuitry must be added to this system to provide for ac/dc rectification, output signal amplification, set point control, and the many other features we look for in sophisticated control systems.

The current through R1 and R2 is

$$I = \frac{E}{R} = \frac{10}{10 + 20} = \frac{10}{30} = \frac{1}{3} \text{ ampere}$$

The voltage drop across R1 is

$$E = IR = \frac{1}{3} \times 10 = 3\frac{1}{3} \text{ volts}$$

and the potential at point A is

$$E = 10 \text{ volts} - 3\frac{1}{3} \text{ volts} = 6\frac{2}{3} \text{ volts}$$

As we look at R3 and R4, we note that nothing has changed so that the potential at point B is still 5 volts. Notice that there is a difference in potential from point A to point B of $6\frac{2}{3} - 5 = 1\frac{2}{3}$ volts. The bridge is unbalanced; the voltmeter will read $1\frac{2}{3}$ volts and we now have the potential to do work. The work done is to open a valve or close a damper or produce some other action that will cause a change in the HVAC system.

If that were the whole story of electronic systems, we could all become experts easily, but there is more to learn.

Note that with a rise in temperature the potential at point *A* became greater than that at point *B*. Current flow could be established from *A* to *B* through a controlled device that causes motion in one direction. If the temperature fell, however, the potential at *A* could drop below the potential at *B* causing the current flow to reverse (remember, this is dc not ac) reversing the action of the controlled device.

In actual systems the resistances of the sensors are quite high (in the neighborhood of 2000 ohms) and the resistance changes with temperature are quite small (around 100 to 200 ohms). The result is that potentials on the order of 50 to 100 millivolts (0.050 to 0.100 volts) are found across points *A* to *B*. These potentials are too small to do much work and must be amplified, or increased, as much as 100 times. This amplification equipment is one source of complexity in the electronic system.

In addition, the typical power supply is 115 volts ac. This must be changed, or rectified, to 10 volts dc. The rectification circuitry is another source of complication. Finally, if we want adjustable setpoints, temperature compensation, reset ratio, enthalpy control, radio frequency interference protection, and so forth, the complexity increases still more.

Fortunately, the controls person need not be an electronic technician. If he or she knows what desired effect is sought and understands a simple vocabulary unique to electronic controls, is patient, and has a logical mind, much can be accomplished.

In the next two chapters we shall take a closer look at the hardware found out in the field. Gradually, the mystery surrounding such controls should become clearer.

DISCUSSION TOPICS

1. What elements make up an electric circuit?
2. What determines the power supply available for a particular application?
3. Why are higher voltage power supplies used for larger HVAC applications?
4. What size of electrical conductor would you specify for a motor with a nameplate current of 28 amperes at 240 volts ac? For a rating of 8 amps at 120 volts ac?
5. What is the difference electrically between a thermostat and a humidistat?
6. What part of a switch dictates its current rating?
7. What are some differences between a pictorial and a schematic wiring diagram?
8. What is an electric relay?
9. Why must pneumatic control system air be clean and dry?
10. Describe how a direct-acting pneumatic thermostat works.

11. What is throttling range?
12. What is the function of a pneumatic relay?
13. How does the term *positive coefficient of resistance* relate to electronic control systems?
14. What is a Wheatstone bridge? What is the result of an unbalanced bridge?

Chapter Nine   Controlled Devices

In HVAC systems we are most concerned with controlling fluids. The two most important fluids to be controlled are air and water. The flow of air can be controlled by means of fans and dampers, while the corresponding control elements in water circuits are pumps and valves. It is interesting that people who have been brought up in the hydronics business have little trouble transferring this knowledge to airflow systems, while duct designers can pick up the principles of water flow equally readily.

There are sufficient differences to warrant separate study, but they are by no means totally independent topics. The majority of controlled devices discussed here will be for air or water. Fuel oil, natural gas, and steam, to mention only three, are also fluids of interest to the HVAC controls person, but they too follow the principles and use hardware similar to those used for air and water. It is a small additional step to understand them fully.

9.1 WATER FLOW

In the previous chapters systems were described that used hot water to distribute heat and chilled water to distribute a cooling effect (or more precisely, to absorb heat). The ability to control the heating and cooling process was dependent on controlling the water temperature or flow rate or both. We have already seen how water temperature is controlled by adding heat in a hot water generator (boiler) or removing heat in a chiller evaporator (liquid cooler). Now we shall see how pumps and valves are used to aid in the control process.

9.1.1 Pumps

A pump is a device that adds energy to a liquid stream enabling it to move through restrictive pipes, fittings, and control valves. The centrifugal pump is the type most commonly used in HVAC systems. It is available in a wide variety of configurations to meet the varying requirements of pressure, flow, and water temperature.

The ability of a pump to do a particular job is defined by a pump curve of the sort shown in Figure 9–2. Actually, we see a family of

FIGURE 9–1   An "in-line" centrifugal pump commonly used in heating applications for such purposes as perimeter heat and duct coil pumping. As flow requirements increase, base-mounted pumps must be used because of their greater size. (Courtesy of ITT Fluid Handling Division.)

curves for a given pump model. Each curve shows a relationship between the flow rate, measured in gallons per minute (gpm) and the discharge pressure, expressed in feet of water, also called total head. Each curve is for a particular impeller diameter, the impeller being the rotating component that does the pumping. The impeller diameter is specified by the design engineer and written into the pump specification.

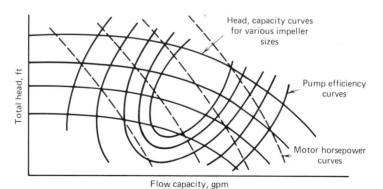

FIGURE 9–2   In selecting a pump/motor combination from the pump curves, it is desirable to optimize the efficiency while meeting pressure and flow requirements.

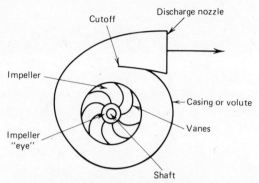

FIGURE 9–3   In the commonly used centrifugal pump, the water is drawn into the eye of the impeller. Centrifugal force created by the rapidly rotating impeller forces the water outward to the discharge nozzle.

Also shown on the pump curve is a series of motor horsepower curves for a number of standard motors that will provide the required energy to produce specified pressure and flow. From the viewpoint of the controls person the controlled device is the pump *motor* rather than the pump itself. The pump control system is designed to energize an electric motor of particular electrical characteristics. The control system may be electrical, pneumatic, or electronic, but at some point it will energize an electric motor with a power requirement of 115/1/60 or 230/1/60 up to about ¾ horsepower or 208/3/60 in larger sizes. (Recall from Chapter 8 that the abbreviated way of specifying electrical power supplies is voltage/phase/frequency.)

The pumps are designed for constant speed operation so that the switching used to energize the pump motor is two-position, on-off. We shall discuss the characteristics of these motors as well as the switching used shortly. It turns out that both pumps and fans use similar control schemes; therefore, we shall consider them together.

9.1.2 Valves   The valves used in controlling water flow automatically are commonly of two types, two-way and three-way. This is true of the small "hold-in-the-palm-of-your-hand" types used in small fan coil units or the larger sizes used in central plants. The two-way valve is simple to apply. It has an opening, or port, that the water enters and a port that it leaves (two openings, hence two-way valve). The valve can be used in a two-position application or it can be used to throttle or modulate flow. Modulating two-way valves are quite common in steam systems but not quite as common in water systems.

Three-way valves are somewhat more complex. There are three openings on such valves, hence the term *three-way*. If the valve is designed to have two of the ports act as inlets and the third as an outlet, it is called a "mixing" valve. If one port is an inlet and the other two are outlets, it is called a "diverting" valve. The valves cannot be used interchangeably because of their internal construction. Although they look identical externally, the valve manufacturer describes them as "three-way mixing" or "three-way diverting" valves and provides a different model number for each.

Three-way valves can be two-position valves causing all flow to go in one of two directions or can be modulating valves. It is the modulating three-way valve that is most commonly found in HVAC piped systems.

In a control system containing automatic valves, we find that the controlled device is actually the *valve operator*. The valve operator is the component that causes the valve stem to rise and fall. Unlike pumps that usually use only electric motors, the valve operator is generally powered by the same power source used in the control system, that is, electric, pneumatic, or electronic power. Since the dampers used in airhandling systems are similarly powered, the discussion of valve operators will be included with damper operators.

One such operator can be described here because it is unique to valves. This is the solenoid. Solenoid valves are commonly available for gas and liquid application up to about 1 inch in size. The solenoid is a length of wire, many hundreds of feet long in some cases, wrapped around a hollow core. A plunger or valve stem is housed within the core. When an electrical potential is imposed across the coiled wire, a magnetic force is created tending to rapidly lift the plunger or force it downward, opening or closing the valve. The force generated by the electrical coil or solenoid is fairly small, tending to limit applications to small-size valves. Another characteristic tending to limit size is the rapid opening and closing that takes place. The valve tends to slam open or closed in less than a second. In a large, high pressure water system such sudden opening and closing can cause an unacceptable hammering noise and fatigue stresses. Solenoid valves are very valuable and many millions are in use, but not in the larger water valves found in HVAC systems.

Valve Flow Characteristics

Although it is the job of the design engineer to fully specify a control valve, one aspect of this specification is of particular interest to the controls person, namely the flow characteristics of the valve. This flow characteristic can be defined as the relationship between the flow rate in gallons per minute (gpm) through the valve and the travel of the valve stem as it moves from fully closed to fully open. At first glance it might be thought that when the valve stem is halfway through full travel, the flow will be 50% of maximum and, when three quarters of the way up, the flow will be 75% of maximum. This is a logical assumption but turns out to be true for only one very special flow characteristic called "linear."

The flow characteristic is directly related to the shape of the plug used in the valve. In Figure 9–4 we see some of the more commonly available plug designs. Each will have a unique flow characteristic. Incidentally, one should keep in mind that these are "control" valves, not the gate or globe valves used for manually shutting a water system off. In Figure 9–5 some typical flow characteristics are plotted on a graph. Note the "quick-opening" valve that establishes almost 70% flow when the stem has traveled only 40% of full stroke. Compare that to the throttle plug which must travel through all of 60% of its stroke to get only 40% of full flow established.

The equal percentage characteristic is interesting. It turns out that the heat output of a hot water heating coil used in a duct is not directly related to the water flow rate. A rather small flow of water produces a large heat output, and increasing the flow rate beyond that initial small flow produces only relatively small increments of additional heat. Figure 9–6 illustrates this heat output versus flow rate situation. A valve with a linear characteristic would not be a good choice for this hot water coil although it would be a good choice for steam heat where heat output is directly related to steam weight flow.

The equal percentage plug was designed to open rather slowly initially and more quickly toward the end of its travel. Note that a 60% travel produces only about a 20% flow rate. However, this 20% flow rate produces about a 60% heat output in a hot water coil. The term *equal percentage* then refers to travel versus heat output rather than travel versus water flow rate.

It is worthwhile for the controls person to be aware that large valve travel may not always produce a large effect on the HVAC system. The performance of the system depends on both the flow characteristic of the valve *and* the action of the valve operator. When confronted with a perplexing problem, it is a good idea to review the valve specification and see what kind of situation you are confronting. On new jobs and on older, well-maintained systems such information should be on file. When such information is not available, as much information as is available from the valve nameplates can be jotted down before calling the manufacturer for help.

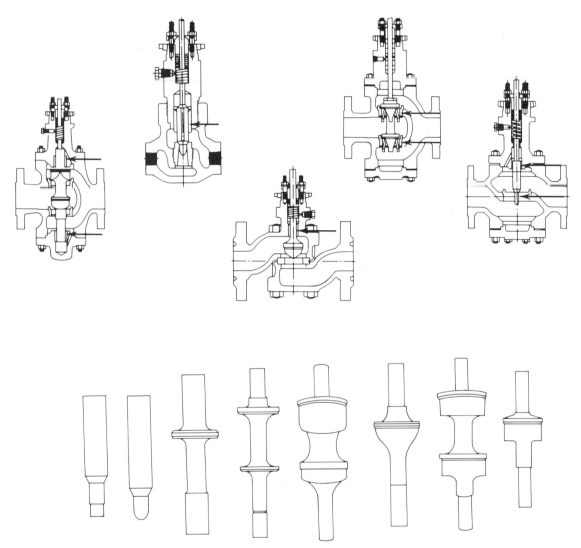

FIGURE 9–4   Several designs of two-way valves with differing plug configurations. Included is a close-up view of the manner in which plugs are shaped to produce desired flow characteristics.

9.2 AIRFLOW   As already stated, airflow control is important to the effectiveness of the HVAC installation. Such control is achieved by means of fans to move the air and dampers to regulate and direct the flow.

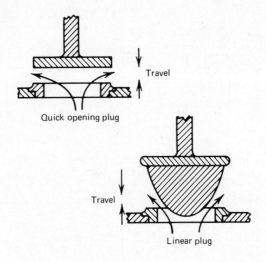

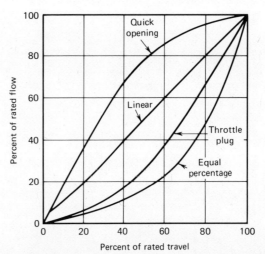

FIGURE 9–5   Several flow versus percentage of valve travel curves. The ability of a valve to control flow is dependent on its plug contours.

9.2.1 Fans   There are two broad categories of fans of interest to the HVAC controls person, centrifugal and axial flow or propeller fans. Although fan design is a science in itself, the application of fans is fairly straightforward. Centrifugal fans are used in duct systems because they are more efficient and quieter than axial flow fans when moving air against a resistance. A forward-curved centrifugal fan is commonly found on small packaged ac units, fan coils, and the like. Larger systems employ either backward-curved or airfoil blades. Other blade configurations may be used from time to time,

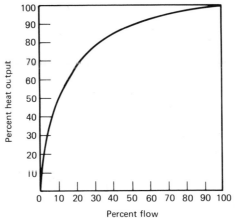

FIGURE 9-6   The heat output of a hot water heating coil plotted against the flow of hot water passing through the coil. Note that a valve passing only half its rated flow still delivers 90% of the heat to be extracted by the duct coil.

but these are the most common. Propeller fans are used when large amounts of air are to be moved against little or no resistance. Relief and exhaust applications commonly employ such fans. These fans are quite noisy compared to centrifugals and must be used where such noise can be tolerated.

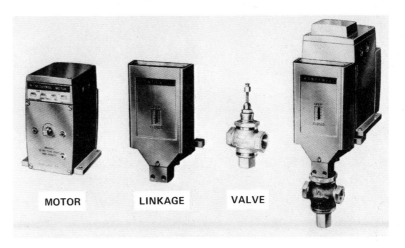

FIGURE 9-7   The motorized valve is actually made up of three components: the motor, the linkage that converts the rotational motion of the motor to the linear motion required to open and close the valve, and the valve itself. Note that the position of the valve can be quickly determined by viewing the indicator on the face of the linkage assembly. (Courtesy of Honeywell Inc.)

The fan curve shown in Figure 9–8 is comparable to the pump curve shown previously. It enables the design engineer to select the proper fan and motor combination to move a particular volume of air, measured in cubic feet per minute against a particular resistance, measured in inches of water column (in. W.C.). Although most pumps use direct drive motors, and therefore operate at one speed (rpm), many, if not most fans are belt-driven, and depending upon the pulley ratio, can be operated at any of a number of speeds. Variation in the output of a pump is accomplished by using different-size impellers, while a similar variation in fan output is accomplished by adjusting the drive-to-driven pulley ratio. We therefore find a family of curves for a fan just as we did for a pump.

As with a pump, the controlled device of the fan control system is the fan motor. It is driven by the standard electrical power sources, yet the control system itself can be electrical, pneumatic, or electronic. Control is typically "two-position, on-off" although some smaller motors are beginning to use solid state controllers that provide infinite speed control in response to a control variable. Variable air volume systems do vary airflow but use dampers rather than fan speed control to do this. Basically, the fan motor is either energized or deenergized in the same manner as pump motors.

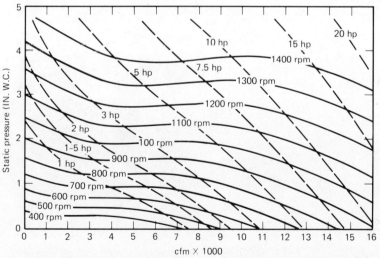

FIGURE 9–8   Fan curves for a 15-inch, forward, curved centrifugal fan. The motor horsepower increments are for available motors. The rpm increments can be set by the proper selection of drive and driven pulley ratios. When a desired flow versus pressure point falls between two motor horsepower ratings, the next larger motor is selected.

9.2.2 Dampers   Controlling the volume and/or direction of airflow is accomplished with dampers. The automatic dampers in use today generally fall into one of two categories, parallel blade or opposed blade. Both are depicted in Figure 9–9.

The parallel blade damper is commonly used in open-closed applications. Fresh air intake and exhaust dampers are very often of this type. Opposed blade dampers, because of their mixing characteristics at low pressure drop are usually used in applications requiring close tolerance flow control. Balancing dampers, zone dampers, return air dampers, and face and bypass dampers are typical applications.

The operator attached to the damper is the controlled device. It may be electrical, electronic, or pneumatic, and it may be two-position or modulating.

9.3 DRIVE MOTORS   The drive motor causes the impeller in the pump to turn, moving water or causing the fan to rotate, moving the air. With very few exceptions it will be an ac electric motor. The exceptions are those pumps and fans driven by dc power, steam turbines, hydraulic motors, windmills, and other unusual power supplies that will not be covered here.

The ac electric motor requires a power supply to make it run and a switching mechanism to turn it on and off. The switching

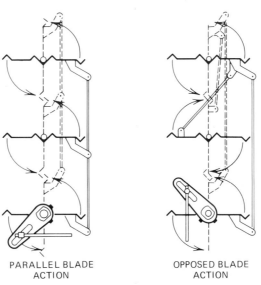

PARALLEL BLADE
ACTION

OPPOSED BLADE
ACTION

FIGURE 9–9   Multiblade dampers of the parallel or opposed type are most commonly used in duct applications to control airflow accurately.

mechanism is the domain of the controls person whereas the power supply is usually the domain of the electrician. The operating characteristics of a drive motor can usually be obtained from the nameplate of the motor.

9.3.1 Nameplate Data

Figure 9–10 shows an example of a motor nameplate. The amount of information given varies from one manufacturer to another. Some common items of information follow.

Frame. A number assigned by NEMA (National Electrical Manufacturers Association) is the frame number, which describes certain physical characteristics of the motor such as the shaft diameter, mounting dimensions, height of the shaft above the base, and similar information. Usually, motors having the same frame number will fit in the same location. Their electrical characteristics may not be the same, but they are physically interchangeable.

Horsepower. An indication of the power available from the motor is called the horsepower. When using fan or pump curves, the brake horsepower selected for a particular application must be equal to, or less than, the nameplate horsepower.

Volts/Phase/Frequency. The power requirement of the motor is described in volts/phase/frequency. Incoming voltage must be within 10% of the nameplate volts. The frequency is beyond the control of the operator and requires special measuring equipment. This is not usually a concern. The phase may be single- or three-phase. The motor will require one or the other and cannot be rewired to accept both.

RPM. The speed of the motor is measured in revolutions per minute at full load. When replacing motors, it is important that the

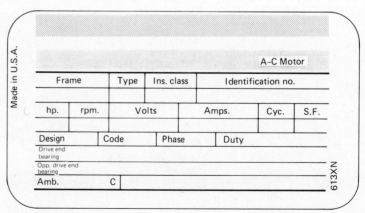

FIGURE 9–10   This nameplate is one of any number of designs likely to be encountered in the field. Different manufacturers will provide what they believe is pertinent information on the nameplate.

RPM of the new motor match that of the old; otherwise, load requirements will change, affecting the system performance and possibly affecting the life of the new motor.

Thermal Protection. It is common for smaller motors to be protected internally from excessive heat. Larger motors usually require external protection built into the control circuit. If the nameplate does not indicate that the motor is protected, one must assume that it is not and provide external protection in the switching circuit (to be discussed further shortly).

Amperes (Amps). Amperes are sometimes listed as FLA or *full load amperes.* This is the maximum current draw for the motor when fully loaded. In belt-drive fan applications the pulleys can be adjusted to increase airflow until the actual current draw matches the FLA. Any additional adjustment may overload the motor leading to early failure.

Service Factor. A measure of the amount of overload the motor can tolerate at rated voltage is called the service factor. Multiply this number by the FLA and get the SFA (Service Factor Amps), which is the maximum allowable current draw. In the absence of a listed SF assume it to be *one;* then the SFA equals the FLA. Service factors as high as 1.35 can be commonly found. As a general rule the higher the SF, the better the quality of the motor.

FIGURE 9–11   A nameplate that includes a wiring diagram showing how the motor should be wired to operate at 230 volts (low voltage) or 460 volts (high voltage). Note also the current draw of the motor at each voltage (9.6 amperes at 230 volts and 4.8 amperes at 460 volts).

Ambient Temperature. The maximum temperature to be permitted to surround the motor is called the ambient temperature. It is commonly expressed in °C (Celsius). At higher temperatures, motor life will be shortened. In fact, for every 10° above the maximum allowable ambient temperature, the life of the motor is halved.

9.3.2 Motor Drives

A direct-drive motor is one that connects directly to the load. The load, fan blade or pump impeller, turns at the same speed as the motor. In some cases the load is mounted directly on the motor shaft, commonly done with smaller axial fans. In other cases a coupling connects the motor shaft to the shaft of the load, commonly done with larger pumps. The term *close coupled* is used to describe motor/pump combinations where the pump housing is mounted directly onto the end of the motor.

In fan applications it is not unusual to see a multispeed motor in use. Occasionally, a selector switch is used to manually select the desired speed. In some cases only one or two different speeds are in use and perhaps another one or two speeds are not used, with

(a)

FIGURE 9–12   (a) An example of a close coupled pump, and (b) a motor/pump combination joined by a coupling. (Courtesy of ITT Fluid Handling Division.)

(b)

their motor leads being taped off. Small furnaces and air conditioning airhandlers up to perhaps 5-tons capacity may use such a configuration.

Belt-drive fans are quite common on larger installations. The number of belts increases from one up to six or more as the horsepower of the motor and required airflow increases. In order for power transmission from the motor to the fan to be accomplished efficiently, all belts should be properly tightened. Belts that are too loose result in power loss; belts that are too tight cause motor and fan bearing wear and, in extreme cases, may wear out the pulley as well.

9.3.3 Motor Switching

The simplest way to energize a motor is to manually flip a switch. From an earlier discussion of electrical circuits we recall that the switch must be capable of carrying the current needed by the motor. The larger the motor and the greater the current draw, the heavier the switching required. In a 120-volt single-phase motor circuit there is only one hot leg; therefore, the switch needs to break only one leg (the hot one). This can be a "single-pole" switch. In a 230-volt single-phase circuit, which has two hot wires, it is best to break both legs so that a "double-pole" switch is used; for three-phase power with three hot legs a three-pole switch is used.

If the motor is small enough, a small toggle switch may be used. Larger motors are usually equipped with larger *disconnect switches,* either fused or unfused, as previously shown in Figure 8–3. In some applications manual switching is adequate, but more often than not automatic control systems require automatic switching.

The device used to accomplish such automatic switching is called a *relay* if the current draw is rather small, say up to about 18 amperes, or a *contactor* if the current draw is higher, and a *motor starter* if motor protection is also included in the switching mechanism.

All three devices use the same principle of operation. A voltage imposed across a coil of wire will produce a magnetic effect. This effect can be used to open or close electrical contacts in the same way a manual switch would. This is the same as the solenoid valve operator described previously. The relays and contactors are quite similar in that they have a coil and one or more switches built into the same body. The coil is electrically independent of the switches. The required voltages of the coils most frequently used may be 24, 120, 230, or 208 volts. The switches, however, are usually designed to take high voltages, perhaps as high as 600 volts. Of course, they will handle the lower voltages commonly encountered with no difficulty.

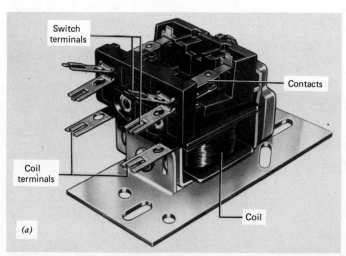

FIGURE 9–13   (a) An open type relay. Note the coil and the exposed contacts. Such relays are commonly found in equipment control panels. (b) An encapsulated relay identical in function but protected by its housing. It may be found in panels but can also be used in exposed locations mounted on junction boxes. (Courtesy of Honeywell Inc.)

(b)

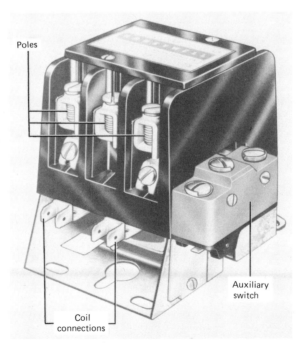

Poles

Coil
connections

Auxiliary
switch

FIGURE 9–14   A contactor that has three poles plus an auxiliary switch mounted at the
right. The auxiliary switch has a much lower rating than the main poles
and is used to energize indicator lights or to interlock with control circuits
of other loads in the system. (Courtesy of Honeywell Inc.)

Highlights—Add-on Cooling

One reason for the popularity of warm air heating systems in small residential applications is their adaptability to add-on cooling at some future time. The wiring diagram in Figure 1 shows the controls for a heating-only system. By replacing the heating thermostat with a heating/cooling thermostat, the single-speed fan motor with a two-speed motor, and by using a fan center such as that shown in Figure 2 the control system will be appropriate for cooling as well as heating. The wiring diagram in Figure 3 shows how the modified system will work.

SEQUENCE OF OPERATION

Call for Cooling
1. Power from the transformer flows through terminal R of the thermostat to both Y and G.
2. Current from Y energizes the cooling unit contactor located in the remote condenser.
3. Current from G energizes the coil of the fan relay, which closes the switch to the high speed winding of the fan motor. The system is now operating in the cooling mode.

Call for Heating
1. Power from the transformer flows through terminal R of the thermostat to W.
2. Current from W energizes the heating control (gas valve or oil primary control) causing ignition.
3. Buildup of heat in the plenum causes the temperature-actuated switch to close, energizing the low speed winding of the fan motor. The system is now operating in the heating mode.

There are any number of ways a system can be modified for both heating and cooling. The wiring diagram was provided by Honeywell Inc. and indicates application of their controls. Whenever attempting any modification of control systems be sure to read all the manufacturer's instructions carefully, particularly *footnotes* and *caution* notes.

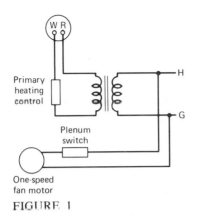

Primary heating control

Plenum switch

One-speed fan motor

FIGURE 1

FIGURE 2

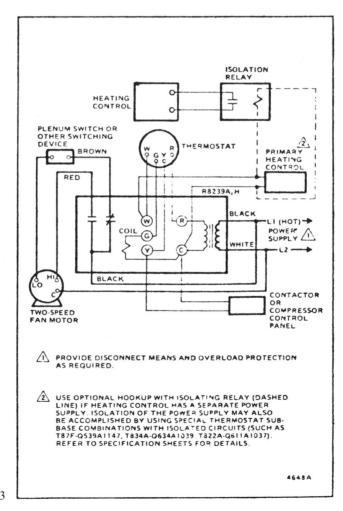

△1 PROVIDE DISCONNECT MEANS AND OVERLOAD PROTECTION AS REQUIRED.

△2 USE OPTIONAL HOOKUP WITH ISOLATING RELAY (DASHED LINE) IF HEATING CONTROL HAS A SEPARATE POWER SUPPLY. ISOLATION OF THE POWER SUPPLY MAY ALSO BE ACCOMPLISHED BY USING SPECIAL THERMOSTAT SUB-BASE COMBINATIONS WITH ISOLATED CIRCUITS (SUCH AS T87F-Q539A1147, T834A-Q634A1039, T822A-Q611A1037). REFER TO SPECIFICATION SHEETS FOR DETAILS.

4648A

FIGURE 3

In summary at this point, if we can impose a voltage across the relay or contactor coil equal to the coil rating, the switches (contacts) will close and the pump or fan motor will begin to operate. By reducing that voltage to zero, the switches will open and the motor will stop.

The job of energizing the coil of the relay or contactor is performed by the control system whether it is electrical, pneumatic, or electronic.

In Figure 9–15 the contactor coil is located in a 24-volt control circuit. If the pilot contact is closed, the coil will be energized and the switches labeled *C* will close energizing motor *M*. The pilot contact is an electrical switch. If the main control system were pneumatic, this pilot contact would be called a P-E switch (pneumatic-electric). It would close when the pneumatic signal reached a specified level, say 8 psig. If the main control system were electronic, this would be the contact of an electronic relay and it would close when the electronic signal reached a specified level, say 6 volts dc. If the system were electrical, the pilot contact might be the switch of a relay in another circuit or a thermostat or pressurestat contact.

The motor starter is a contactor with motor protection added. The discussion in the previous section is perfectly applicable to motor starters. Not all motors have built-in protection. The motor

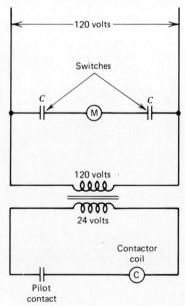

FIGURE 9–15   Although separated in the schematic diagram, the coil and switches labeled *C* are contained in the same piece of hardware.

FIGURE 9–16   This P-E (pneumatic-electric) switch is used to start an electric motor in response to a pneumatic signal. A variety of switching arrangements are available to provide flexibility in control system design. (Courtesy of Johnson Controls Inc.)

starter provides protection to the motor in the event that it begins to draw excessive current due to some sort of overload.

In Figure 9–18 the schematic of a motor circuit with a motor starter is shown. In series with the coil and pilot contact are two normally closed switches labeled *S*. They are overload switches responsive to heat. Notice in the lines going to the motor the two

FIGURE 9–17   Several motor starters with selector switches labeled "hand-off-auto." Note the plate identifying each one, Pump 5, Fan TX-1, and Fan TX-2.

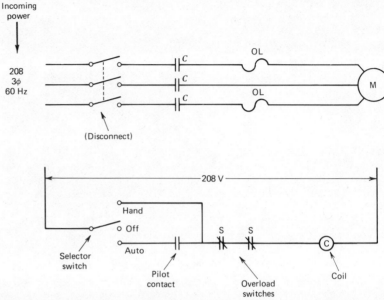

FIGURE 9–18   The control circuit energizing the coil of the motor starter can be any appropriate voltage, in this case 208 volts, the same as the motor (M). In this system the motor starter has three contacts or poles. If the manual disconnect is in the closed position, with the selector in the auto position, the motor will respond to the pilot contact, which could be a thermostat switch, pressure switch, or even a contact in another relay.

heaters labeled *OL.* If the current to the motor becomes excessive, the heaters warm up and generate heat. This heat is felt by the overload switches and if they get too hot they will open, deenergize the coil, and shut off the motor. Generally, there is a manual reset feature built into these controls so that if the overload shuts off the motor, a button must be pushed by the building operator to get it going again.

The heaters are designed to generate a controlled amount of heat dependent upon the current draw of the motor. A particular motor starter can accommodate a number of different-size heaters. The heaters must be carefully selected to provide adequate protection yet not trip out prematurely. A heater selection chart accompanies each motor starter; the chart is usually pasted on the inside cover of the motor starter box.

Although automatic control systems generally require automatic switching of the motors, in some cases a manual push button arrangement is desirable. It is not unusual to have the supply fan motor of a large system started with a pushbutton arrangement, which in turn energizes other control circuits through *interlocking controls.* The other circuits will automatically bring on line such equipment as return air fan motors, pumps, and compressors.

An increasing use of motor starters with a ''hand-off-auto'' selecter switch is taking place. The ''hand'' position is a manual switch that allows the operator to bypass any pilot contacts and energize the motor independently of any other controls in the system. The ''off'' position permits service to be performed on the motor without shutting down the entire system. The ''auto'' position is the normal operating mode for fully automatic operation.

9.4 ACTUATORS

Actuators are devices that cause valves and dampers to open and close. Usually, such motion occurs slowly in response to a control signal generated by controllers. The major exception to this is the solenoid operator, previously described, that causes quick opening and closing. The control signal can be electrical, electronic, or pneumatic.

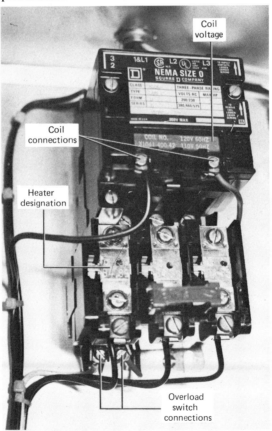

FIGURE 9-19   The coil of a motor starter with a rating of 120 volts. This coil can be readily replaced with one of a different voltage if necessary. Note the heater designation of B4.85. A check of the manufacturer's installation instructions shows that this is the heater to be selected for a motor with a current draw between 3.16 and 3.66 amperes.

Highlights—Part Winding Starting

The current draw of a motor at the instant of start can be as high as five to six times the normal running current. This high starting current can cause voltage fluctuations in the line that may affect other equipment connected to the same power source. Dimming lights, radio and television interference, and interference with data processing equipment are several results of high inrush current. In some areas power companies have required that measures be taken to reduce such effects.

One way of reducing starting current is to use motors and controls designed for part winding starting. These are quite common in HVAC equipment with motors larger than 10 horsepower. The wiring diagram in Figure 1 is for a 25-horsepower air conditioning compressor motor. Note that six wires are connected to the compressor, three from contactor C1 and three from contactor C2. The object of the controls shown is to energize part of the motor windings initially (terminals 1, 2, and 3) and the rest of the motor windings (terminals 7, 8, and 9) from 1 to 3 seconds later.

Contactor C1 is energized on a call for cooling by the thermostat. When its contacts pull in, they energize part of the compressor windings *and* a time delay relay. The time delay relay, after an appropriate period of time (usually about 1 second) will energize contactor C2 and the compressor will reach full speed. Although the inrush current is still substantially greater than the normal running current, it is significantly lower than that achieved with "across the line" full voltage starting using a single motor starter.

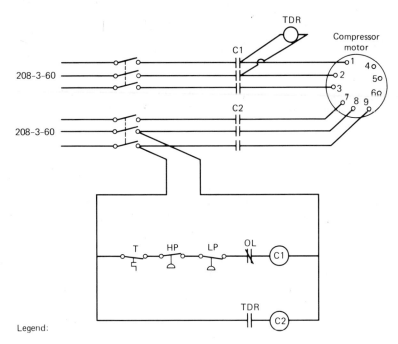

Legend:

    T = Thermostat
   HP = High pressure cutout
   LP = Low pressure cutout
   OL = Compressor overload
   TDR = Time delay relay
   C1 = Compressor contactor—part winding
   C2 = Compressor contactor—part winding

FIGURE 1

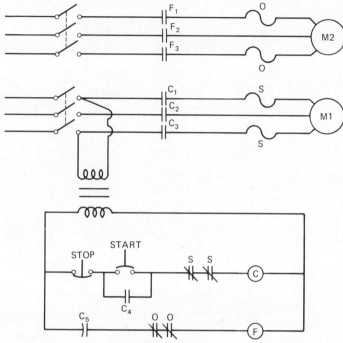

FIGURE 9–20   Motor starter C has five poles in this system. Three control the motor; one, labeled C_4 is a bypass around the manual start button to keep the coil energized even when the button is released; and the fifth, labeled C_5, is an interlock to motor starter F. Motor M2 cannot start before motor M1, in this arrangement; hence it is "interlocked."

9.4.1 Electrical Actuators

An electrical power supply of 24, 120, or 208 volts may be used to drive these actuators. These motors move slowly, perhaps ½ to 1 revolution per minute, and usually do not rotate more than 180° as the damper or valve goes from full open to full closed.

The two types of motors available are two-position or modulating. Two-position motors are available as reversing and spring return. A reversing motor is one in which the motor windings are energized to cause it to rotate one way and back again. A spring return motor is one that must be energized to rotate fully one way. It remains energized at the extreme position. When deenergized, a helical spring, which had been wound by the rotation of the motor, now unwinds and causes the motor to return to its original position.

An electrical diagram for a spring return motor is shown in Figure 9–22. The power source may be any of the popular supply voltages. The control contact may be the contacts of a thermostat, humidistat, or other control. When the contact closes, the motor winding from X to Y is energized causing the motor to rotate

(a)

(b)

FIGURE 9–21   Two types of reversible modulating motors available in voltages compatible with line and low voltage control systems. Rotational speeds are on the order of 1½ rpm. Note the warning that the shaft should not be turned with a wrench. It's very difficult to turn and any effort will usually result in damage to the motor. (Courtesy of Honeywell Inc.)

against the helical spring. As the motor shaft rotates, a cam, located on the shaft, also rotates. When the motor has rotated fully, say 160°, the bump on the cam hits the limit switch causing it to energize the motor winding from *Y* to *Z*. Under this condition just enough torque is generated by the motor to overcome the load and to keep the spring wound. As soon as the control contact opens, the motor is deenergized and the spring torque causes the motor to move back 160° to its original position.

One type of reversing motor is shown in Figure 9–23. The electrical wiring is such that when voltage is imposed across *A* and *C*, the motor will rotate in one direction. When voltage is imposed across *B* and *C*, the motor will rotate in the reverse direction. The

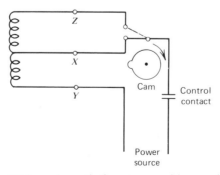

FIGURE 9–22   Wiring schematic for a two-position, spring return motor.

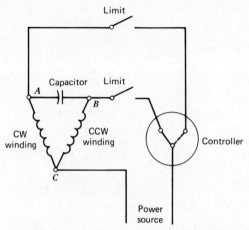

FIGURE 9–23   Note the capacitor in this reversible induction motor. This motor will rotate in a clockwise direction with the controller in the position shown. A cam on the motor shaft will trip the limit switch, deenergizing the motor after it has gone full travel.

control circuitry of this motor must include limit switches that deenergize the motor winding after the valve or damper actuator has gone through its full stroke. If this is not done, a locked rotor condition will exist and the motor may burn out. In addition, the linkage connecting the motor to the controlled device must be accurately adjusted to insure that the valve or damper is at its extreme open or closed position at the time the limit switch opens. Linkage adjustment is very important, yet not adequately understood by many people in the field. A good number of motors are burned out on new installations because of improperly adjusted linkages.

One other characteristic of some two-position reversing motor control circuits is that they must permit the motor to move fully to the open position once it has started to open, and fully to the closed position once it has started to close. Even if the controller attempts to reverse the motor direction in midstroke, the motor will complete its stroke before reversing.

The proportional motor uses a bridge circuit, quite similar to the one described, for electronic controls. Figure 9–25 gives a simplified schematic of one such proportional motor. The major components of the system are the modulating controller, balance relay, reversible induction motor, and feedback potentiometer. In the diagram shown the motor and control circuit use the same 24-volt power source. Similar systems are available using a motor powered by 120 or 240 volts. The control circuit (controller, balance relay, and feedback potentiometer), however, still uses 24 volts. In some cases a 24-volt transformer is built into the motor; in

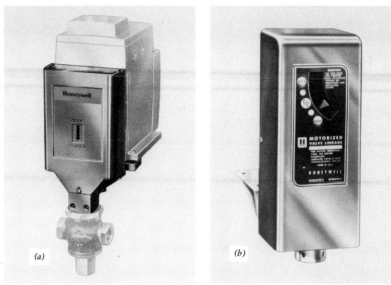

FIGURE 9–24   Two linkage assemblies. A linkage is required to connect the actuator to the valve or damper. Careful adjustment of the linkage is necessary to prevent the electric actuator from going into a "locked rotor" mode and possibly burning out. Strict instructions accompanying the linkage mechanism should be carefully followed at the time of installation. (Courtesy of Honeywell Inc.)

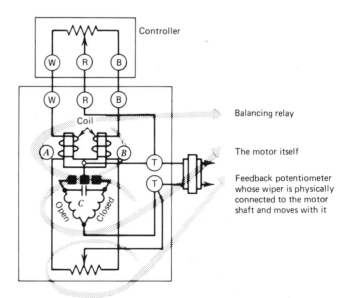

FIGURE 9–25   This modulating motor permits dampers and valves to be positioned at some intermediate point in their travel proportional to the heating or cooling load on the system.

others an externally mounted transformer is required for the control circuit only. The principle of operation is the same, but the internal wiring configurations are different.

The principle of operation is rather simple once the function of the balance relay is understood. Notice that the relay has two legs, each having a coil of wire surrounding it. When current flows through the legs of the relay, a magnetic force is created that acts on these legs. When the current through each is the same, the switch contact is at an intermediate position and neither winding of the reversing motor is energized. The motor is stopped. If the current in leg A exceeds that in leg B, the magnetic force acting on leg A overcomes that acting on leg B and the relay switch contact completes the circuit to winding C causing the motor to begin turning. Two questions should arise at this point. First, why should the current in the two legs be different? Second, what prevents the motor from moving full stroke once it begins to turn?

In Figure 9–26 the schematic has been simplified still further to show the controller, the two legs of the balancing relay, and the feedback potentiometer. The controller also is in the form of a

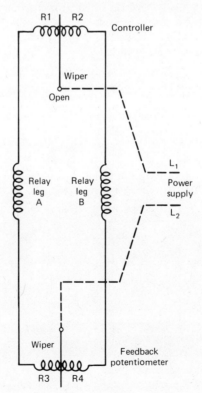

FIGURE 2–26   Power can be thought to be entering the upper wiper, flowing in two directions through the relay legs, and leaving at the lower wiper.

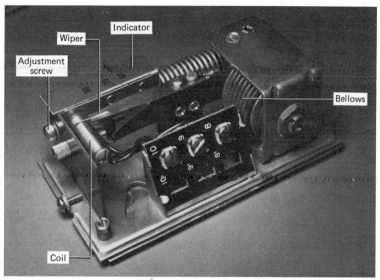

FIGURE 9–27   The inner workings of a proportional controller. A change in temperature is felt by the bellows, which expands or contracts causing the potentiometer wiper to move across the potentiometer coil. This unbalances the bridge circuitry causing the proportional motor to begin moving. Note the adjustment screw for selecting the setpoint, and the scale on which the setpoint is indicated.

potentiometer, a coil of high resistance wire with a wiper arm that rides across it. The wiper has the effect of creating two resistances, hence R1 and R2 in the controller potentiometer, and R3 and R4 in the feedback potentiometer. Although the controller may be at a location quite remote from the motor, the three wires connecting the two have very low resistance and the controller location is usually not a factor in the operation of the motor.

The 24-volt power supply is imposed across L1 and L2, and two circuits are energized: R1, relay leg *A*, R3; and R2, relay leg *B*, R4. Here R1, R2, R3, and R4 are variable resistances depending upon the position of the potentiometer wiper. The controller wiper is positioned by the room temperature and the feedback wiper by the motor shaft rotation.

Let us look at this arrangement in operation. We assume a starting point of 70°F room temperature and both the controller and feedback potentiometer wipers are at midpoint so that R1 = R2, and R3 = R4, current through leg *A* = current through leg *B*, and both motor windings are deenergized. A rise in temperature to 71°F causes the control potentiometer wiper to move to the left. The resistance of circuit R1, leg *A*, R3 decreases and that of R2, leg *B*, R4 increases due to this movement. The current through leg *A* of the relay increases; the current through leg *B* decreases; and the

FIGURE 9–28   A proportional motor that responds to the proportional thermostat. As the shaft turns, the feedback potentiometer wiper moves over the potentiometer coil rebalancing the bridge and bringing the motor to a halt. Note the two auxiliary switches mounted next to the shaft. They can be used to interlock other control devices and can be made to open or close when the shaft rotates to a selected position by the cams mounted on the shaft. A typical use of such switches would be to shut off a compressor after a set of multizone dampers have gone to the full heating position, or perhaps to energize a signal light when a set of dampers is open.

relay contacts move, energizing winding *C* and causing the motor to rotate. That should answer the first of the two questions.

As the motor shaft rotates, the feedback wiper moves to increase R3. It will continue to do so until the total resistance of circuit R3, leg *A*, R1 is once again equal to that of R2, leg *B*, R4, at which point current through both of these circuits will be equal, the relay contacts will open, and the motor will stop turning. In the event that the temperature change sensed by the controller is so great that the two circuits never get back into balance, the motor will continually be driven in one direction and if this situation is left unchecked the motor will burn out. Limit switches are built in, much the same as in the two-position reversing motor, to deenergize the motor when it is at the end of its travel even though the control circuit is not rebalanced. This then is the answer to the second question.

9.4.2 Electronic Actuators

Electronic control systems typically produce signals between 6 and 18 volts dc after extensive amplification. Such a potential is rather small to open large sets of dampers or control valves used in high pressure gas or water systems. Generally, the electronic actuators employ common ac voltages, such as 24, 120, and 240 volts, to produce the required force but use the lower dc voltages to control the higher voltage input.

One such electronic actuator receives the dc signal from the bridge circuitry described earlier. This signal is sensed by solid state components, which in turn control the 120 volts ac power going to a reversible induction motor of the type used in electrical actuators. As the motor rotates, a feedback potentiometer mounted on the shaft moves to rebalance the bridge in a manner similar to that already described. The solid state controls do not have open contacts like the balancing relay and limit switches of the electrical actuator. Their function, however, is identical and their external appearance is quite similar. Proportional, two-position, and two-position with spring return configurations are available as electronic actuators.

Another type of electronic actuator uses an electrohydraulic principle. In Figure 9–29 we can see a small hydraulic pump, typically powered by line voltage. The discharge of the pump goes to the pressure chamber. The pressure level in this chamber is determined by the position of the dump valve, which is controlled

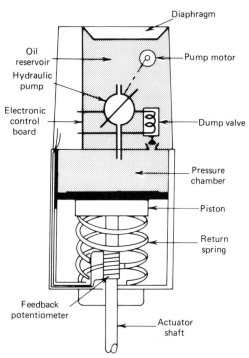

FIGURE 9–29   The voltage to the hydraulic pump of this electronic actuator can be any appropriate line or low voltage. The electronic signal is connected to the dump valve.

by the dc control signal of the bridge circuit. If the signal causes the valve to close, pressure will build up and the actuator shaft will extend. As the valve opens more, pressure will decrease and the shaft will retract. Note the feedback potentiometer on the shaft. Movement of the shaft varies the resistance tending to rebalance the bridge circuit, achieving proportional control. Two-position control can be obtained by designing the dump valve to either fully open or close dependent upon the magnitude of the dc signal.

It is not uncommon to see electronic and electrical actuators both used in a large system. A two-position electrical actuator can be energized and deenergized by relays having dc coils. The dc signal opens or closes the switch in the ac power line going to the motor. Although new installations usually use one or the other system, older renovation jobs may use both.

Just a quick word about troubleshooting such systems. Solid state electronics (and electronics in general) is a science all by itself. It is frankly asking a lot for a person to be well versed in HVAC systems *and* in electronics. Fortunately, this is not necessary. The descriptions given here are quite sketchy purposely. To

FIGURE 9–30   This electro-hydraulic actuator positions air dampers or valves in HVAC systems. An important aspect of any electronic component is that it be compatible with the controller, main panel, and other electronic components in the control system. (Courtesy of Johnson Controls Inc.)

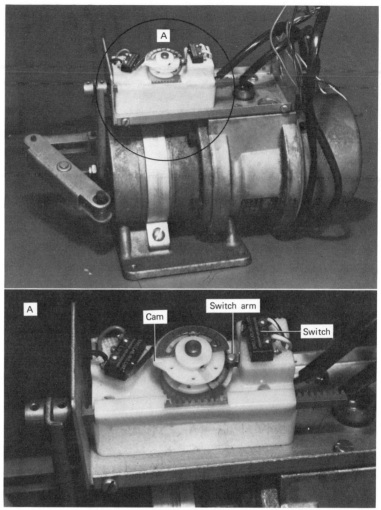

FIGURE 9–31   An electronic actuator that is also of the electrohydraulic type. It is equipped with end switches or auxiliary switches responsive to the position of the shaft. Note the cams that rotate as the shaft is extended or retracted. They can be set to trip the auxiliary switches at any desired shaft position.

troubleshoot these systems, one *must* have the manufacturer's manuals and procedures. We shall look at some of these procedures in detail later. They are easy to follow and require a minimum knowledge of electronic circuitry.

9.4.3 Pneumatic Actuators

Pneumatic actuators were discussed in Chapter 8. They respond to either low control pressures in the range of 3 to 15 or 18 psig and, where larger forces are needed, up to 100 psig.

FIGURE 9–32   A pneumatic actuator mounted on a valve. Note the wide diameter of the diaphragm at the top of the actuator that permits a large force to be obtained with a low control pressure. (Courtesy of Johnson Controls Inc.)

Pneumatic actuators are inherently proportional in that the signal from the controller, passed directly to the actuator, will produce a proportional response. Two-position control can be obtained but usually requires the addition of a pneumatic relay. The term *relay* was used in electrical systems to describe an electrically operated switch. In pneumatic controls the term *relay* is given to a device that alters in some way the pneumatic signal coming from the controller. A relay would receive a gradually increasing signal and at a predetermined point, say a pressure of 8 psig, it would allow a 15-psig signal, that is, full pressure, to go to the pneumatic actuator causing it to move from a fully retracted to a fully extended position.

In some systems electrical and pneumatic controls are used together. An E-P relay, actually a solenoid valve, might be energized allowing full pressure to flow to a pneumatic actuator, causing, as an example, a set of outdoor dampers to go from a full-closed to a full-open position. This is another example of two-position control.

Actuators, although designed to act over a range of pressure from 3 to 15 psig, can be obtained with a more restricted operating range within those limits. It might be desirable to sequence a series of pneumatic actuators. As the controller generates a gradually increasing signal, one actuator may go from a fully retracted

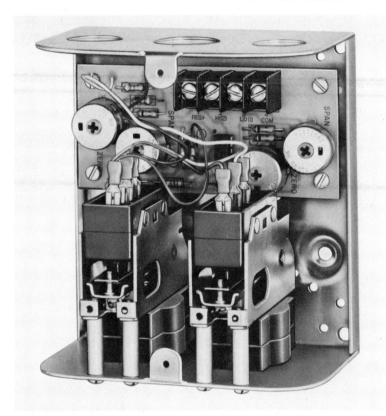

FIGURE 9–33   Pneumatic actuators can be energized by electronic control systems through the use of "transducers" or "interface" devices. Electronic wiring is connected to the terminal board at the top and pneumatic tubing is connected to the controllers at the bottom. (Courtesy of Johnson Controls Inc.)

position to a fully extended position as the pressure changes from 3 to 8 psig; another begins to react at 8 psig and is fully extended at 12 psig; and a third only begins to move at 12 psig and is fully extended at 15 psig. Another arrangement may call for one actuator, which is fully extended at 3 psig, to retract fully as the signal reaches 8 psig; a second remains fully retracted from 3 to 10 psig but then begins to extend at 10 and is fully extended at 15 psig. Generally, the action and the pressures are specified at the time of purchase and are set at the factory rather than in the field. The appropriate selection of relays and actuator operation gives pneumatic systems a tremendous amount of flexibility.

At this point we have seen how changes in the HVAC system performance can be accomplished. In the next chapter the way in which signals are generated as a result of control variable changes will be examined.

**DISCUSSION
TOPICS**

1. How are pumps and fans similar and how are they different?

2. How does a solenoid valve differ from a motorized valve?

3. How would you describe the flow characteristics of a quick-opening control valve compared to an equal percentage control valve?

4. What fan speed would you select to produce an airflow of 8500 cfm against a static pressure of 3 in. W.C.? What motor size would you select? Would the motor be fully loaded? (See Figure 9–6.)

5. Of what use are motor nameplate data?

6. What is the difference between a disconnect switch and a contactor? Between a contactor and a motor starter?

7. Give an example of how you might apply the principle of interlocking controls to an air conditioning system.

8. Why would a spring return damper motor be used on a fresh air intake damper?

9. What would happen if the limit switches in a proportional reversing motor failed to open when desired?

10. What is a potentiometer?

Chapter Ten   Controllers

The controller senses a control variable such as temperature or pressure, compares its value with a desired set point, and generates a signal to a controlled device. The controlled device reacts to that signal and causes corrective action to be taken to bring the actual control variable value back to the desired value. In the previous chapter we looked at a number of controlled devices. In the following pages we shall look at how the signal to which they respond is generated.

10.1 TEMPERATURE CONTROL

There are more devices concerned with maintaining a desired temperature in an HVAC system than with any other function. The two major categories of such devices are thermostats and temperature controllers. There is no very good reason to regard them separately because they work in very similar ways. The major difference is that generally a thermostat is located in the conditioned space and senses the temperature of the air surrounding it. The temperature controller senses the temperature at a remote location and transmits a signal some distance to the apparatus that analyzes the signal and generates a control signal. The signal from a thermostat or a controller is the same, and the controlled device does not know whether it is being controlled by a thermostat or a controller.

10.1.1 Sensing Temperature

In electronic control systems the temperature is sensed by means of a wire element whose resistance varies as the temperature changes. As an example, one particular electronic sensor has a resistance of 1772 ohms at a temperature of 74°F. It has a positive coefficient of resistance, which means that as the temperature increases its resistance will also increase. In addition, it has a sensitivity of 4.2 ohms per degree, which means that as the temperature rises, say 1°F to 75°F, the resistance will increase by 4.2 ohms to 1776.2 ohms. Sensing temperature electronically, then, is nothing more than being able to determine the resistance of a wire element, and, knowing the properties of the wire convert that resistance reading in ohms to temperature units. The

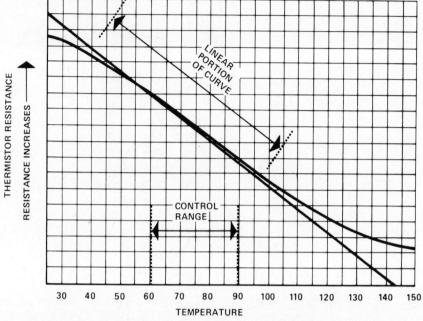

FIGURE 10–1   The curve for a thermistor having a negative coefficient of resistance. Note that the decreasing temperature produces an increasing value of resistance. The range of temperatures over which a thermistor is accurate is usually confined to the linear portion of the curve. (Courtesy of Honeywell Inc.)

Wheatstone bridge circuitry can do this. Simple instruments employing thermistors (a fancy name for a temperature-sensing resistor) are widely used in measuring temperatures in service work.

Pneumatic and electrical control systems, much more commonly encountered than electronic systems, use mechanical means of sensing temperature changes. The mechanical sensor converts a temperature change into motion. This motion causes electrical contacts to open or close in the case of electrical two-position control or causes a potentiometer wiper to move across a resistance wire in the case of electrical proportional control; or it can cause a flapper valve to either restrict or release airflow in the case of pneumatic control.

There are in general use today two techniques for creating the motion due to temperature change. One is the use of a bimetal element. The bimetal element consists of two dissimilar metals, each with a different coefficient of thermal expansion, bonded together. As the temperature of the element rises, the two metals will expand at different rates. The element will then distort in a

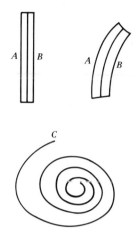

FIGURE 10-2   A leaf-type bimetal element composed of two dissimilar metals, A and B. A temperature rise has caused a deflection of the element toward the right because the coefficient of thermal expansion of A was greater than that of B. Below a helix is formed of a similar bimetal material. A change in temperature will cause point C to move in a clockwise or counterclockwise direction.

way dependent on how it is shaped. A leaf-shaped element will move laterally; a helical shape will tilt; and a longitudinal helix will increase or decrease in length. The control designer must determine which motion is wanted and how he or she will harness this motion to produce a control signal.

Another technique for producing motion as a result of temperature change is the pressure system. A fluid (gas or liquid) contained in a closed vessel will increase in pressure on a temperature rise and decrease in pressure on a temperature fall. The relationship between temperature and pressure is predictable and dependent upon the fluid used. Some controls are liquid-filled; some are gas-filled; and still others use a combination of liquid and vapor.

In thermostats a bellows arrangement is frequently used. As the temperature rises and falls, the bellows expands and contracts producing an axial motion. The Bourdon tube, a curved tube of elliptical cross section is used not only in controls but also in gages. A change in temperature produces a change in pressure, which causes the end of the tube to move. Another scheme employs a diaphragm. Temperature change produces pressure change which causes the diaphragm to deflect.

Whatever the arrangement used, the motion must eventually result in a control signal being generated by the temperature controller. A mechanical linkage transmits and often amplifies the motion. In a two-position controller this motion causes electrical contacts to move together or apart. In a modulating electrical

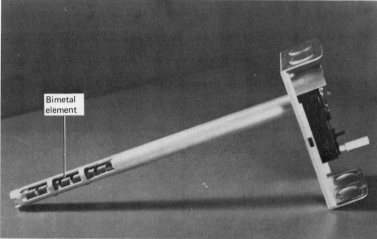

FIGURE 10–3   Two examples of the use of bimetal elements to sense temperature. The wall-mounted thermostat at the top uses a helically wound bimetal that twists a mercury switch into an open or closed position as the temperature changes. The combination fan/limit switch at the bottom uses a longitudinal helix, inserted into the supply airstream of warm air furnace to sense the temperature.

controller a moving wiper causes the resistance in two circuits to change, producing an imbalance in a relay connected to a reversible induction motor. In a modulating pneumatic controller, the motion causes a flapper to seek a new position relative to a nozzle, causing a particular pressure to occur in the form of a signal to a controlled device.

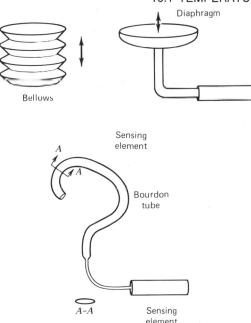

FIGURE 10–4   The bellows, diaphragm, and Bourdon tube, examples of closed pressure systems that sense a temperature change and cause a motion to take place. In some applications the bellows alone senses the temperature; in others a capillary may connect it to a sensing element such as that shown attached to the diaphragm.

10.1.2 Detent Action

Changes in the air temperature of an HVAC system are rather slow. Assume that two electrical contacts are joined making a circuit. A very slow change in temperature acting on a bimetal element will result in a very slow drawing apart of the two contacts. There is a minimum air gap across which electricity will jump. In a 24-volt system the gap is very small while in 120- and 208-volt systems the gap is increasingly large. Electricity jumps the gap in the form of a high temperature spark, which we call "arcing." Arcing is a normal occurrence and rarely will an electric switch open or close without some small spark. Arcing for a prolonged period causes electrical contacts to pit, char, and eventually burn out. Controls for two-position application are designed to minimize arcing by having detent action built in.

Detent action is the quick-opening and quick-closing effect that occurs when the set point is reached. It is accomplished in a number of ways. Perhaps the most common example is the glass encapsulated mercury switch used very widely in residential thermostats. The helical bimetal element sensing temperature causes the capsule to tilt one way or the other sending the small mass of mercury in the desired direction to make or break the circuit. The

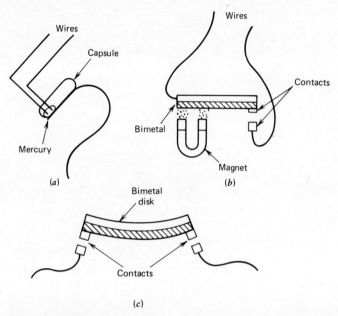

FIGURE 10–5   Three arrangements of detent action. *(a)* The encapsulated mercury, *(b)* the magnetically attracted bimetal, and *(c)* the "oilcanning" bimetal disc all tend to open and close the electrical contacts rapidly to minimize the effects of arcing.

FIGURE 10–6   This steam pressure control illustrates one type of detent action. As the pressure approaches the set point of the control, the capsule containing mercury will tilt downward at the left causing the mercury to flow away from the electrical contacts opening the circuit to the steam boiler burner unit.

mercury moves as a unit very quickly and arcing is of very short duration. Incidentally, on any number of occasions, homeowners have called for service because their thermostat was sparking or "on fire." Actually, what had happened was that they had walked by their thermostat in the dark just as it happened to be switching and noticed the arcing. They normally would not see it in the daylight and rarely would they be looking at it when it switched, in any event; hence their concern when they accidentally saw it arc.

Another method of detent action uses a magnet. As the bimetal, sensing the temperature change, moves one contact toward the other, it gets within the magnetic field of a small magnet and the contact is quickly pulled shut. On opposite temperature change a stress builds up in the bimetal until it overcomes the magnetic force and quickly snaps open.

"Oilcanning," commonly used in snap action temperature and current overloads in compressors, and "over-center" toggle action used in pressure controls are two other examples of detent action.

10.2 CONTROLLER CHARACTERISTICS

In selecting a temperature controller, it is well to be aware of some characteristics that may vary from one to another.

Operating Range. The range of temperatures over which the controller functions is the operating range. A room thermostat may have a range of 55 to 90°F, whereas a high temperature controller may have a 240 to 385°F range. Obviously, the desired set point must be within the operating range of the control. A good practice is to select a control operating range such that the desired set point is within the middle third of the range.

Differential. The differential is a characteristic of two-position controls that indicates the difference between the cut-in and cut-out temperature of the controller. In room thermostats it might be 1 to 1½°F degrees whereas in controllers used in other applications it might be 5, 10, or even 20°F. In some controls the differential is set at the factory and cannot be changed in the field, but others are field-adjustable. This is an important characteristic for good temperature control.

Maximum Operating Temperature. The highest temperature the control can sense and still function reliably is the maximum operating temperature. It is usually associated with controllers employing remote sensors. If the temperature being sensed will rise to 200°F, it is apparent that a controller with a 125°F maximum operating temperature is inappropriate.

Electrical Ratings. The electrical contacts of the controller are designed to handle a particular voltage/current combination. A room thermostat may only be able to handle 1.5 amperes at 24 volts, whereas a controller in a unit heater may be rated at 8 amperes at 120 volts, a much higher electrical rating. A switch will not last long if subjected to higher electrical loads than that for which it was designed.

Throttling Range.   The throttling range, which applies to proportional controls, is the number of degrees change required to drive the controlled device from a minimum to a maximum position. In some cases it is fixed at the factory; in others it is field-adjustable. One typical controller has a fixed throttling range of 3°F, which will provide rather a rapid response to system changes. Another controller of similar design has a throttling range that is field-adjustable from 9 to 39°F, which provides a degree of flexibility in setting up a system to enable the service person or operating personnel to minimize *hunting* (another term for system cycling or swinging around the set point).

Capillary Length.   In remote controllers, where a sensing bulb is connected to the body of the controller by a thin tube called a capillary tube, the capillary tube length varies from as little as 5 feet to as much as 20 or 30 feet. Capillary tube material, which is usually of copper, is sometimes provided in monel or stainless steel for high temperature or corrosive atmosphere applications.

Switching.   Switching refers to the internal wiring of the controller. Perhaps most common is the SPST (single-pole, single-throw)

FIGURE 10–7   Although somewhat similar in appearance the two controllers shown here have markedly different characteristics. The device at the left has a standard sensing bulb, a range of 0 to 100° F, a SPST switching action, a 1-ampere switch rating at 120 volts ac, and a 1° F fixed differential. The device at the right has a fast response element, a range of 160 to 260° F, a SPDT switch rated at 1 ampere at 120 volts ac, and a 3.6 to 12°F adjustable differential. Some information is obvious by examination, but other data must be obtained from the control specification sheet. (Courtesy of Honeywell Inc.)

(a)

(b)

FIGURE 10–8   (Left): a line voltage control thermostat used to directly energize unit heater blower motors, water pumps, and similar line voltage loads. (Right): a low voltage control thermostat used for heating/cooling applications. Note that both controls incorporate a thermometer so that the set point and actual temperature can be compared. (Courtesy of Honeywell Inc.)

switch for turning a system on or off. Some controllers use SPDT (single-pole, double-throw) switching to shut off one system and turn on another as the temperature changes. Sequential switching or staging may be included in the specification of switching action along with a statement of the delay in degrees between the first and second stage being energized.

Mounting.   Some controls are designed for wall mounting, some for electrical junction box mounting, some for attachment to an immersion well, some for use in an explosive atmosphere, and so forth. The physical characteristics of controls cannot be overemphasized. The device must fit in the allocated space and be able to be installed conveniently.

Sensing Element.   In liquid-sensing applications an immersion well or perhaps a surface mounting scheme is used. A thermostat is installed in the conditioned space. A controller may have a plain capillary tube with or without a bulb on the end. The sensing element may be quick-responding, spot-sensing or averaging. The proper control application depends upon the proper selection of this feature. The specification sheet for the controller will provide all the information needed to identify the function of the device. With some experience the reader of specification sheets will be able to select the information that the reader needs. The controls designer, installer, troubleshooter, and operator each has requirements for information somewhat different than the others.

T636A CROP–TROL CONTROLLER

USED FOR SPACE TEM-PERATURE CONTROL OF ROOM AIR CON-DITIONING UNITS OR RADIATOR VALVES.

If used with changeover switch or thermostat, these devices can control both heating and cooling. Sensing bulb installed in return airflow. One light duty spdt switch. Range: 55 to 90 F [13 to 32 C]. Differential: 1-1/2 F [0.8 C], fixed. Bulb Size: 5/16 x 11-11/16 in. [7.9 x 296.9 mm]. Capillary: 5-1/2 ft [1.7 m]. Temperature Scale Marked: Warmer-Cooler. Case Dimensions (L6018A only): 5-5/8 in. [142.9 mm] high, 2 in. [50.8 mm] wide, 2-1/4 in. [57.2 mm] deep. Listed by Underwriters Laboratories Inc.—L6018A; component recognized by Underwriters Laboratories Inc.—L6018E;

ELECTRICAL RATINGS (amperes):

	120V ac	240V ac
Full Load	8.0	5.1
Locked Rotor	48.0	30.6
Millivoltage	0.25 amp at ¼ to 12V dc.	

T636 SWITCHING.

FIGURE 10–9   An excerpt from a control catalog illustrating the type of information that must be known in order to properly apply a control to a particular field situation. (Courtesy of Honeywell Inc.)

10.2.1 Controller Action   We have seen how a change in temperature is translated into a mechanical motion. This motion produces different results depending on the type of thermostat. On a heating thermostat we would expect a *fall* in temperature to cause a set of contacts to close completing an electrical circuit to the controlled device in the heating system, perhaps a gas valve or an oil burner. Such a switch would be an SPST switch (single-pole single-throw) and is the least expensive, most widely used thermostat. In a cooling thermostat the requirement would be that on a *rise* in temperature the SPST switch would close completing the circuit that would energize the cooling equipment.

Increasingly systems are being used that have a heating and cooling capability. A single thermostat employing an SPDT (single-throw, double-pole) switch is used. A problem with this switch is that it makes one circuit immediately upon breaking the other. Applied directly as SPDT switch, this thermostat would energize the cooling equipment immediately after it shut down the heating equipment and would turn on the heat as soon as it shut down the cooling unit. This is obviously unacceptable. The

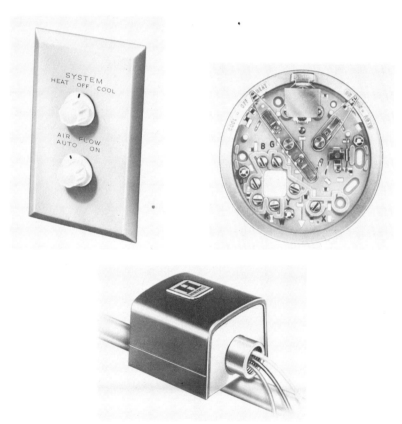

FIGURE 10–10   Three control methods of changing over a system from heating to cooling. The system switch shown at the upper left and the thermostat subbase at the upper right provide for manually switching over. The control at the bottom is mounted on a water line and will switch over a fan coil control, for example, based on the temperature of the water in the line. (Courtesy of Honeywell Inc.)

changeover switch is added to prevent this. In some thermostats it is a selector switch enabling a "cool" or "heat" mode to be selected. When in the cool mode the heating unit cannot be energized even though the SPDT switch of the thermostat demands it. To increase the flexibility of the basic thermostat, a number of switching functions, including the manual changeover, are built into a "subbase." The subbase contains manual switching for fan operation, heating, cooling, and other auxiliary equipment.

In more sophisticated systems, usually larger commercial applications, the changeover may be accomplished automatically external to the main thermostat. In some applications a thermostat mounted outdoors might accomplish it, or a thermostat mounted on a water line sensing hot or chilled water and thereby correcting the action of the room thermostat might accomplish changeover to provide heating or cooling as appropriate. Finally, a manual switch located in a central plant control room may change over the action of any number of thermostats located throughout a building.

A floating controller arrangement is gaining in popularity as the concern for energy consumption increases. In this control, changeover is accomplished by the controller in the space. If we assume that the controller is calling for cooling, as the temperature falls the cooling system will be deenergized. At this point the system will "float"; that is, neither heating nor cooling will take place. The range of temperature over which neither heating nor cooling equipment is energized is called the "dead band" and may be as great as 4 or 5 degrees. A continued drop in temperature will eventually call for the heating system to come on. Maximum advantage can be taken of the building's ability to store the heating and cooling effect and to use solar heating by passive means.

In a proportional electrical thermostat the change in temperature results in a potentiometer wiper's moving across a resistance wire. The incoming power (24, 120, or 208 volts) flows through the wiper and then in two directions through the resistance wire. In Figure 10–11 the incoming power enters R and flows out of B and W. (Although this is alternating current, the explanation is simplified by assuming current flow in one direction only.) You may want to refer back to Figure 9–25, which shows the controlled device used with a proportional controller. The moving wiper is changing the resistance in each leg of the balancing relay circuit to cause the reversing motor to move one way or the other. The thermostat is designed to be part of a specific control system and must be compatible with the controlled device. This is not necessarily true of two-position thermostats, where you can use just about any manufacturer's control to do a job assuming that its electrical rating is adequate.

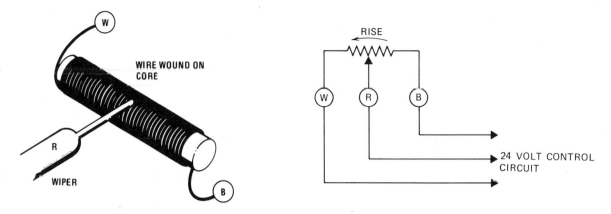

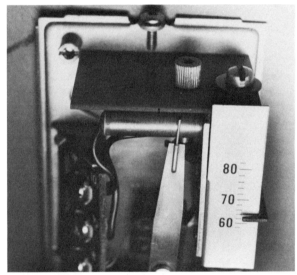

FIGURE 10–11   A diagram typical of those found in manufacturers' literature. In this case the switching of a proportional controller is shown. The photograph shows the actual components that provide the switching depicted in the diagram. (Courtesy of Honeywell Inc.)

These controls can be readily checked with an ohmmeter. In two-position thermostats continuity through a switch is checked. A zero reading indicates a closed switch; an infinite reading indicates an open switch. In checking a proportional controller the manufacturer's specification sheet must be available to indicate the actual resistance reading in ohms that are necessary when the control is in a prescribed position. This is a bit more involved but certainly not beyond the ability of anyone reading this book. In some cases directions for calibrating a proportional control thermostat call for the system to be energized and require voltage readings at the control. Again, by following the step-by-step instructions and the safety directives provided by the manufacturer, one should have no trouble.

Troubleshooting controllers is a combination of art and science. Some words of advice at this time are appropriate. Be patient. Be careful. Follow all safety directives contained in manufacturers' instructions. Read all instructions *completely* before attempting any troubleshooting procedures. When confused, the troubleshooter should walk away, take a break, and think of something else for a few minutes, then return and start at the beginning.

10.2.2 Control Location

There are as many, if not more, problems caused by improper control location than by control malfunction. The control responds to the temperature felt by its sensor. The sensor is usually quite small, perhaps only an inch or two long, but the area being serviced by the HVAC system may be thousands of cubic feet in volume. Locating the sensor where it will feel a sample of the air that is representative of this vast volume becomes very critical.

A room thermostat should be mounted about 60 inches above the floor on an interior wall or column. The height is about right for people walking around an office or for a living space. The interior surface location ensures that hot or cold exterior walls do not influence the sensor, which is supposed to be feeling the air temperature.

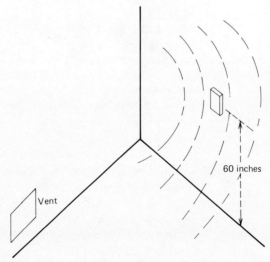

FIGURE 10–12   A wall-mounted thermostat, which should be mounted about 60 inches off the floor in a location that senses air that is representative of the air in the conditioned space. It should be out of the stream of air blowing from ceiling or sidewall supply outlets.

SENSING ELEMENTS

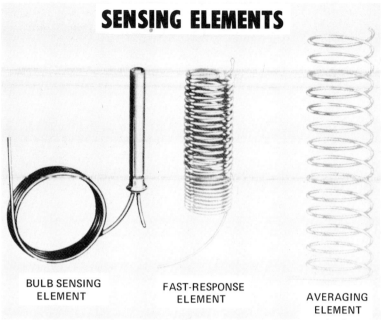

BULB SENSING
ELEMENT

FAST-RESPONSE
ELEMENT

AVERAGING
ELEMENT

FIGURE 10–13   A bulb sensor that can accurately reflect the temperature at only one point. If this one point cannot be selected so that it is close to the average of the space, perhaps the averaging bulb ought to be selected. If a rapid response to changing conditions is desired, a quick acting sensing element would be beneficial. (Courtesy of Honeywell, Inc.)

We should try to place the thermostat where it will sense typical air in the space. A deadend hall or a broom closet does not meet this requirement. Placing it where the opening and closing of exterior doors cause drafts to affect the sensor is also not a good idea. In a ducted system, placing the thermostat as close to the return duct as possible gives relatively good results.

Controllers designed to sense the outdoor temperature should be mounted on the north-facing wall of the building so that solar radiation does not cause them to sense a temperature higher than the actual air temperature. On more than one occasion a controls person has ignored the directive on the engineering drawing to mount a sensor on a north wall and mounted it on the south wall because it was more convenient to do so. The result was a totally ineffective system.

A controller with an averaging bulb strung across a duct will do a much better job of minimizing the effects of stratification than one with a short 2-inch immersion stem.

Some thoughts about controls and their location follow. People like to have control of their environment. In an office with a number of people and even in the home, there will be much changing of the thermostat setting. In commercial locations thermostats are often installed with covers that let air in but keep busy fingers out. Controllers are also provided with tamper-proof mechanisms that require a special key to be used to change the setting. Invariably, when there is an equipment malfunction, the first reaction of the equipment operator is to adjust the thermostat. When that fails, the repair person is usually called. The repair person should always check the thermostat setting prior to leaving an installation. A responsible person on the premises should always be contacted and shown how the controls operate, where they are located, and how they can be adjusted. It is important to emphasize that adjustment should not be necessary to minimize the amount of adjustments that will be made by the equipment operators.

One should be especially watchful of controls that operate zone dampers at remote locations. A zone thermostat may be operating dampers that control the air flowing into another zone while another zone thermostat may be placed on the ceiling tiles outside of its zone. When starting new jobs, it is important to look at the whole system and see that all controls not only work but are connected in accordance with the building control plan. One should be careful on old installations or renovation work as well; someone may have placed a thermostat in an out of the way corner without checking it out to see what it does and where it should best be located.

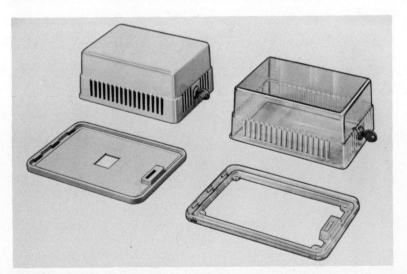

FIGURE 10–14 Protective covers of this type are helpful in discouraging occupants of the conditioned space from making thermostat adjustments. None is totally effective against the really dedicated "dial twiddler," but the presence of such a cover will give warning that adjustments are not encouraged. (Courtesy of Honeywell Inc.)

10.2.3 Control Wiring   A major benefit of 24-volt wiring is that it is light and easy to install and need not be enclosed in a protective raceway. One can touch the live leads of a 24-volt control circuit and not feel a thing. The major shortcoming is that 24 volts is not much of a potential to do work; therefore, it is generally used to energize relays that control line voltage power going to a controlled device. Other than the mechanics of running the wire, which is best left to the electrician, the connections of line voltage and low voltage control systems are quite similar. A word of caution is necessary, though: One should continually check the power source and system components to make sure they are compatible. Nothing is more embarrassing on energizing a system than to have sizzling, smoking control components and motors.

Control manufacturers are aware that control circuit wiring is a major source of problems in the field and they have done much to help the controls person. Very detailed instructions are packed with the controls they sell. One difficulty is that many of the controls can be applied in a number of ways. Frequently, the instructions contain wiring diagrams for six or more applications. The question then becomes, which of the diagrams applies to our particular situation? It is important to be able to read and understand wiring diagrams, to know the system components, and to know the sequence of the operations.

Another manufacturer's aid is identifying the control terminals in accordance with a conventional format. One manufacturer uses letters that correspond with common thermostat wire colors. In that case R (red) is connected to the transformer; W (white) goes to the heating circuit; Y (yellow) goes to the cooling equipment; and G (green) is connected to the fan circuit. After installing a few of these systems, it will not be necessary to refer to the diagram. Unfortunately, there is no code convention on an industry-wide basis. Another manufacturer may identify the thermostat terminals based on their function so that the terminal marked V (voltage) goes to the transformer, that marked H (heat) goes to the heating equipment, C (cool) goes to the cooling equipment, and F (fan) is connected to the fan circuit. This, too, is easy enough to remember, but when it is combined with other conventions used by other manufacturers, it is easy to see why control wiring may be a bit tough to grasp particularly for the newly initiated controls person.

10.3 ELECTRONIC CONTROLLERS   The electronic controller is a very complex system. The field of electronics is changing so rapidly that older electronic systems employing vacuum tubes, newer systems with transistorized circuits, and the newest with microprocessor chips are all in use, in

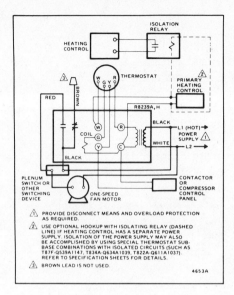

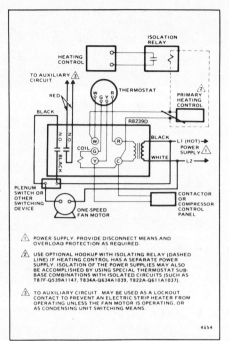

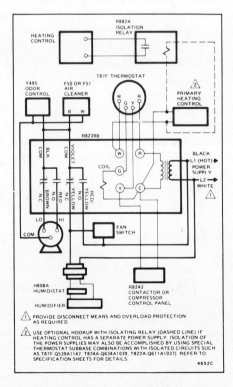

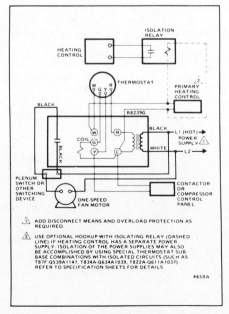

FIGURE 10–15　Four wiring diagrams included in the instructions for a fan center. Although the control is the same, it is being supplied in a number of ways. Each diagram shows a particular application for the control. Care must be taken in selecting the appropriate diagram for your use. (Courtesy of Honeywell Inc.)

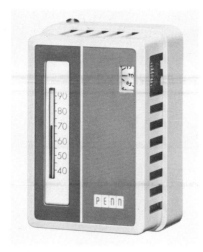 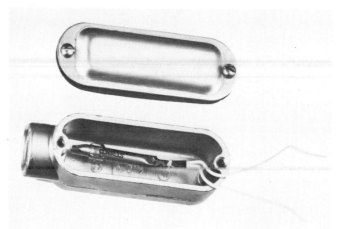

FIGURE 10–16   Two electronic sensors. At the left is a wall-mounted sensor with adjustable set point typically used for zone control. At the right is a nonadjustable sensor used for sensing outdoor temperature or return air temperature. (Courtesy of Johnson Controls Inc.)

some installations side by side. The internal workings of such systems need not be mastered but some appreciation of how they work is useful.

The Wheatstone bridge described previously is the basis for this system. As the temperature rises above the set point, an amplified voltage signal of a particular polarity is generated. The magnitude of this dc signal will vary with the temperature change. If the temperature drops below the set point, an increasing dc signal of opposite polarity is generated. Figure 10–17 gives a typical "voltage ramp" signal from the amplifier. This particular signal is being sensed by a master panel, which in turn controls a single-zone, multistage, heating/cooling central system employing an economy cycle (using a maximum amount of outside air). Let us now examine what happens as the temperature rises and falls.

Notice that at the set point, and for a short span on either side of the set point, a zero-volt dc signal exists (the so-called "dead band"). As the temperature rises to about 0.8°F above the set point, a 4-volt dc cooling signal is generated by the bridge/amplifier circuit of the space controller. The economy cycle begins to function allowing an increasing amount of outdoor air to enter the building providing "free" cooling. Note that the main panel also analyzes a signal from an outdoor thermostat or compares outdoor air and return air enthalpy to determine if the economizer can function or should be overridden. At about 1.4°F above the setpoint, a 6.64-volt dc cooling signal has caused the economizer to be in full operation. A further increase in temperature to 1.8°F above

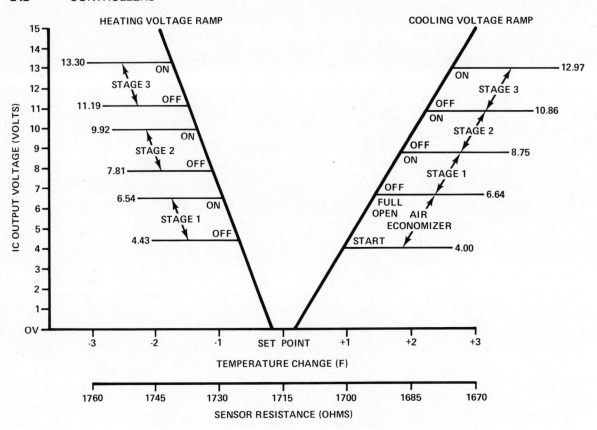

FIGURE 10–17   A typical voltage ramp signal from the Wheatstone bridge of an electronic HVAC control system. This particular system incorporates an economiser cycle and three stages of cooling as well as three stages of heating. (Courtesy of Honeywell Inc.)

the set point results in an 8.75-volt dc cooling signal that energizes an electronic relay to bring on the first stage of mechanical cooling. A continuing rise in temperature brings on the second stage at about 2.2°F above the set point (10.86 volts dc) and a still further increase to 2.6°F above the set point (12.97 volts dc cooling signal) causes the third stage to come on line. The stages could be completely separate units serving a common area, or three compressors in one large unit, or three levels of unloading in a single compressor. If a centrifugal or screw compressor were used, infinite variation of the suction line valve setting could be accomplished using an electronic operator to proportionally adjust the valve in response to the dc signal.

As the temperature in the conditioned space decreases toward the set point, the dc cooling signal would also decrease and the third, second, and first stage of cooling would shut down. The

FIGURE 10–18   The heart of the electronic system is the main panel. Signals from the sensors are analyzed, and corrective signals are sent to electronic actuators at remote locations. At the right are six SPDT relays for switching equipment "on" and "off." Note adjustments such as the outdoor air damper minimum-position potentiometer and outdoor air high limit setting. (Courtesy of Johnson Controls Inc.)

economizer would then be reduced to admitting minimum ventilation air. A similar analysis of the electronic control system can be made as the temperature continues to fall. An increasing dc heating signal (reverse polarity of the cooling signal) will cause the available stages of heating to come on line one after the other. Typical applications use gas valves in parallel, or electric resistance heaters, or hot water coils piped in parallel circuits each having its own control valve.

10.3.1 Servicing Electronic Systems

In this section three sets of instructions for checking out electronic system components will be considered. Note their similarity. A voltmeter capable of reading 0 to 20 volts dc is applied to a set of terminals; one or more dials are turned. If the voltmeter registers the desired response, one can move on to another test. If the required response is not obtained, the component should be replaced.

Operational Checkout of an Electronic Room Thermostat with Integral Power Supply

The following instructions were extracted from an installation data sheet published by the Johnson Controls Company.[1]

The ambient temperature must be within range of the thermostat when this test is performed. Be sure system and supply voltage wiring are correct before power is applied.

1. After the controller is installed and the cover removed, apply power to the controller and associated devices.
2. To check the direct acting output, connect a voltmeter set to read up to 20 VDC, between the red wire (+) and the blue wire (−).
3. Rotate the set point control to the minimum setting, and the D.A. bandwidth to the maximum setting (clockwise). The meter should indicate approximately 16 VDC.
4. Slowly move the set point control toward the maximum setting; the voltage should change from 16 to 0 volts. No other change should take place.
5. To check the reverse acting output, connect the voltmeter between the red wire (+) and the white/blue wire (−).
6. With the set point control and the R.A. bandwidth control both at the maximum setting, the meter should indicate approximately 16 volts.
7. Slowly move the set point control toward the minimum setting; the voltage should drop from 16 to 0 volts. No other change should take place.

If the results of this test are satisfactory, proceed to the Bandwidth Adjustment procedure. If the test indicates faulty operation, check system wiring and associated equipment. If the controller is defective, replace it.

---

[1]With permission of the Johnson Controls Company, Milwaukee, Wisconsin.

Operational Checkout of an Electro-Hydraulic Damper Operator

The following instructions were extracted from an operational checkout and adjustment instruction sheet also published by the Johnson Controls Company.[2]

To check the operation of the actuator, 115 volts or 24 volts A.C. and a variable voltage source of 0 to 16 volts D.C. at 20 ma. must be used. The variable voltage source may be either a power supply or an appropriate controller. If a controller is used, the voltage should be monitored at the test points described in the Operational Checkout and Adjustment Bulletin covering the controller used, and the set point control varied to obtain the voltages required for the following tests. Be sure power wiring and applied voltage are correct before testing.

1. Apply A.C. power to the actuator pump (Terminals 7 and 10).
2. Apply 0 voltage to the servo valve (Terminals 8 and 11).
3. Observe the operation of the actuator. The shaft should be fully retracted.
4. Apply a 16 volt D.C. signal to the servo valve.
5. Observe the operation of the actuator. The shaft should move to the fully extended position, operating smoothly throughout the stroke.
6. Check the operation of the actuator with 10 volts D.C. applied to the servo valve. The actuator shaft should come to rest approximately at mid-point in the stroke as indicated by the graduated scale on the shaft.
7. If the actuator does not operate properly, check the system wiring for possible errors. If necessary, perform resistance checks on the actuator, as indicated in Table I. (See next page.) If the resistance readings differ from those listed in the table, replace the actuator. Return actuators to the factory for repair; do not attempt field servicing.

Cooling Ramp Adjustment of Master Panel Used in Multizone Application

In this test procedure a test instrument (available from the manufacturer) that simulates the signal of a zone thermostat is used to impose a dc signal on the master panel. The procedures pre-

---

[2]Ibid.

TABLE I	PUMP MOTOR	
	115-Volt Motor	24-Volt Motor
Terminal 7 to Terminal 10	11 Ω ± 1 Ω	0.5 Ω (approx.)
Servo Valve	Terminal 8 to Terminal 11 1000 Ω ± 50 Ω	
Feedback Potentiometer	Terminal 5 to Terminal 6 110 Ω ± 10 Ω (Wiper of Potentiometer Connected to Terminal 12)	

sented here are extracted from a Johnson Controls Service Guide and are reproduced here with their permission.

1. Adjust zone temperature simulator for 15 VDC measured between terminals 12 (−) and 14 (+).
2. With 15 VDC high signal, the cooling ramp signal should be 10 VDC. Adjust "Cooling Range" potentiometer for 10 VDC measured between terminals 8 (−) and 9 (+).
3. Adjust zone temperature simulator for 13 VDC measured between terminals 12 (−) and 14 (+).
4. With 13 VDC high signal, the cooling ramp signal should be 4.8 VDC. Adjust "Cool Bandwidth" potentiometer for 4.8 VDC measured between terminals 8 (−) and 9 (+).
5. Repeat Cooling Ramp Adjustment Steps 1 and 2.
6. Adjust zone temperature simulator for 17 VDC measured between terminals 12 (−) and 14 (+).
7. With 17 VDC high signal the cooling ramp signal should be 15 VDC. Adjust "Cooling Bandwidth" potentiometer for 15 VDC measured between terminals 8 (−) and 9 (+).
8. This completes the first approximation for achieving correct cooling ramp. Adjust zone temperature simulator slowly from 12 to 18 VDC while observing the cooling ramp for correct cut-out and cut-in points of cooling stages. If cooling ramp is not in agreement with Figure 1, repeat Steps 1 thru 7 until proper cooling ramp is achieved. Each repetition of Steps 1 thru 7 will successively approach correct cooling ramp.

It is apparent that the key to properly checking electronic systems is having the appropriate manuals and specification sheets for the components to be checked. These are readily available from the manufacturer and actually should be part of the operating manual for the building. It is fairly easy to troubleshoot an *electrical*

FIGURE 1

COOLING RAMP SWITCH OPERATING POINTS

Stage	Cut-In	Cut-Out
1	13 VDC	5 VDC
2	15 VDC	7 VDC
3	17 VDC	9 VDC

system with a wiring diagram and instruments regardless of who made the controls. This is not true of *electronic* systems where much more information is required.

10.4 PNEUMATIC CONTROLLERS

The control signal in a pneumatic system is air pressure in the range of 3 to 15 psig. This pressure signal is generated by a device variously called a transmitter, controller, thermostat, pressurestat, or humidistat depending on the name the particular manufacturer chooses to use and on its function. All of these devices have in common a supply air main with pressures of about 18 psig going to them and a branch carrying the output signal leaving. Generally, the two types of controller we deal with are (1) *direct acting,* where an increase in the control variable produces an increase in the output signal, and (2) *reverse acting,* where an increase in the control variable produces a decreasing signal. The signal itself varies proportionately with the control variable.

Just as in electrical controls the pneumatic controllers can respond to changes in temperature, pressure, humidity, enthalpy, differential pressure, and so forth, and the appropriate controller to do a particular job can be selected from the manufacturer's catalog. A tremendous amount of flexibility can be achieved in the pneumatic control system by using devices called relays to modify the signal from the main controller to achieve certain control actions.

Several kinds of control arrangement are available. Figure 10–21 is a simple scheme employing a wall-mounted thermostat, which, externally at least, is quite similar to electrical and electronic thermostats. Note the air line connections. That labeled M is the main supply pressure, which might be as high as 20 to 25 psig depending on the capability of the controls used. The port labeled B is the branch that carries the control signal, which ranges from 3 to 15 psig, to the hot water valve. The valve is labeled NO for "normally open." This is common for heating valves, since, in the event of a power failure or air pressure outage, the valve will go to a full-open position. This ensures that the building will receive heat, although uncontrolled, should the failure occur in cold

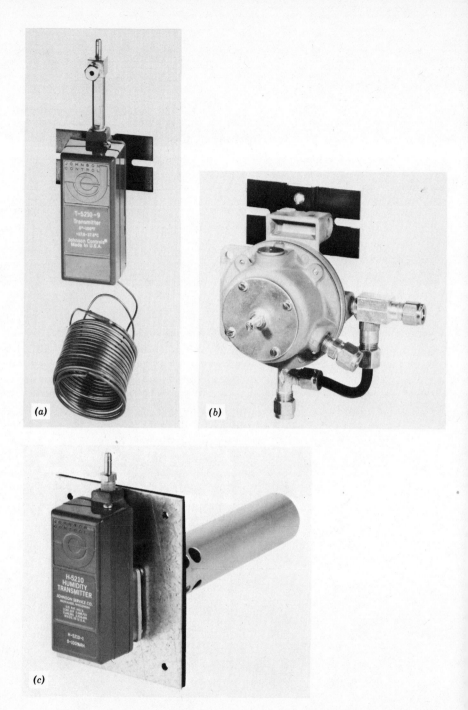

(a)

(b)

(c)

FIGURE 10–19 Devices that sense a variable and generate a pressure signal in the range of 3 to 15 psig dependent upon the value of the variable sensed. This signal in turn goes to a controller that analyzes the signal, comparing it to a set point, and then sends out another pneumatic signal to an actuator to provide corrective action. (Courtesy of Johnson Controls Inc.)

FIGURE 10–20   This controller receives the signal from the transmitter and in turn generates a pneumatic signal in the range of 3 to 15 psig to the damper or valve
actuator. The gage indicates the output pressure signal. The adjustment at
the lower left can control the relationship of the input signal to output
signal (commonly referred to as ''gain''). (Courtesy of Johnson Controls
Inc.)

weather. The thermostat is direct acting (DA) because, as the room
temperature rises, we want the valve to close down. An NO valve
will close with an increase in control pressure. Therefore, a rising
room temperature producing an increasing control signal requires a
DA thermostat.

   In Figure 10–22 the system we just looked at has been expanded.
The thermostat now controls three hot water valves feeding three
heating coils operating in parallel. The object of this arrangement is
to call for heat sequentially as the temperature falls and demand

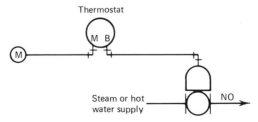

FIGURE 10–21   A simple pneumatic control scheme employing a thermostat and normally
open proportional valve for steam or hot water use.

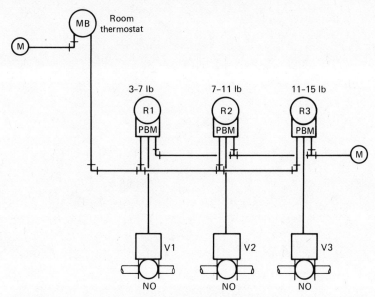

FIGURE 10–22   An example of the flexibility of pneumatic controls in which three separate heating valves are controlled sequentially by a single thermostat providing three-stage operation.

increases. The three valves are identical in that they move from fully closed to fully opened as the pressure signal decreases from 15 to 3 psig. Three pneumatic relays are used to provide the sequential operation. As the room thermostat senses the falling temperature, it will generate a decreasing signal. As the control signal falls from 15 to 11 psig, we want valve V3 to go from full closed to full open; as the control signal falls from 11 to 7 psig, we want valve V2 to open fully; as the control signal finally falls from 7 to 3 psig, we want valve V1 to go from full closed to full open. However, since each valve is designed to go from closed to open as the pressure falls from 15 to 3 psig, the relay, termed by one manufacturer a *pneumatic ratio delay*, must modify the signal from the thermostat so that a 4-psig change in pressure from the thermostat will become a 12-psig change in pressure at the valve actuator. Notice that each relay has an M connection, which means that full system pressure, say 20 psig, is available at the relay. Relay R3 is adjusted so that as the signal from the thermostat felt at port P changes from 15 to 11 psig, the output of the relay at port B changes from 15 to 3 psig. Similarly, relay R2 produces a signal change of 15 to 3 psig when it feels the thermostat signal changing from 11 to 7 psig. Finally, relay R1 gets into the act as the thermostat signal changes from 7 to 3 psig by generating a decreasing signal to valve V2 of 15 down to 3 psig. The sequencing of the valves is thereby accomplished with the first valve V3, then valve

V2, and finally valve V1 going from full closed to full open as the temperature falls. Incidentally, under a partial heating load, V3 may be fully open and V2 may be modulating at an intermediate position. All three valves need not be opened everytime the thermostat senses a temperature change. Such action depends on the magnitude of the change.

A final diagram will give still a better idea of the flexibility of pneumatics. In Figure 10–23 thermostats T1 and T2 are sending signals to relay R1. This relay analyzes the signals and generates a third signal from port B that is the average of the two incoming signals felt at ports P1 and P2. This signal in turn controls the flow of hot water to a heating coil, or the flow of chilled water if needed, to a cooling coil. Actually, the flow of chilled water is being controlled by relay R2 based on signals it is getting from R1 and the humidistat H1. Relay R2 is designed to analyze the incoming signals at its ports P1 and P2, and generate a third signal equal to the *greater* of the two incoming signals through port B. At certain times the zone thermostats T1 and T2 may be satisfied and generate a signal designed to close down the chilled water valve. The humidistat, on the other hand, might simultaneously sense a need

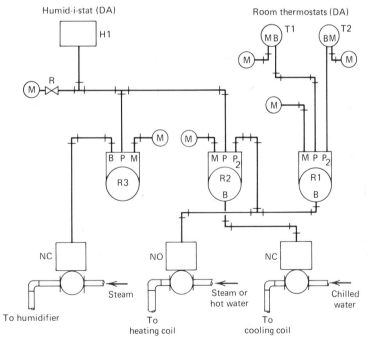

FIGURE 10–23   In this control arrangement heating, cooling, and humidity control can be achieved using pneumatic controllers and relays. The ability of the pneumatic relay to average incoming signals or to select the greater of two incoming signals makes this possible.

for dehumidification and therefore desire that more chilled water flow through the coil. Its signal would be greater than the signal from relay R1 and therefore would be passed by R2 causing the increased chilled water flow with subsequent dehumidification. If this increased cooling effect produced uncomfortably low temperatures in the zones, reheating could be accomplished as the signal from the space thermostats would open the heating valve even though the cooling valve was already open.

The humidistat in this system is direct-acting, as its signal is competing against that of T1 and T2 to control dehumidification. A call for humidification, that is, too low a humidity level in the conditioned space, will result in a low pressure signal from the humidistat. This must open the steam valve used for adding moisture to the air. It is desirable to use a normally closed (NC) steam valve so that in the event of a penumatic system failure, the valve would close preventing the uncontrolled introduction of steam into the HVAC system. Relay R3 is a reversing relay that enables the two conditions just stated to be set. A decreasing signal from the humidistat goes to port P of the relay, where it causes an increasing signal to be generated from port B, which in turn opens the valve.

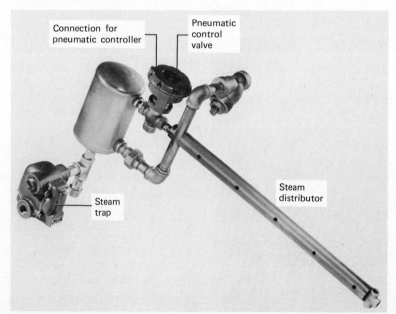

FIGURE 10–24 A steam humidifier employing a pneumatic valve to proportionally control the flow of steam from the distributor tube. Note the steam trap at the lower left to ensure that only condensate, and not steam, returns to the steam generator. (Courtesy of Johnson Controls Inc.)

In this diagram we have seen just three of a number of relay capabilities, namely reversing an incoming signal, selecting the greater of two signals, and averaging two incoming signals. Manufacturers catalogs contain many more.

If the description of Figure 10–23 seems complicated, after reading it over and studying it six or seven times, it gets less complicated and soon it actually seems simple. It is the rare individual who can walk into a building and know how everything works.

10.4.1 Troubleshooting Pneumatic Systems

To troubleshoot pneumatic systems, we must know the sequence of operations, the function of the controls, and how they are intended to respond to particular pressures. We must also be able to read the pressures; guessing will not work. Most larger systems have pressure gages located in every imaginable spot. Some smaller systems have pressure-gage access ports that enable one to attach a service gage. Some controls have a soft pad that can be pierced by a hypodermic needle attached to a service gage for quick access. The pad reseals once the needle is removed.

Before looking at the control system as a source of trouble, it is a good idea to check out the mechanical equipment because, in most cases, the problem will exist there. Also, such equipment can usually be checked out rather quickly compared to the time-consuming effort required for controls. Once the refrigeration compressor, boiler, pumps, and secondary heating and cooling equipment have been eliminated as a source of trouble, we can check the pneumatic system air supply, the air compressor, pressure-reducing stations, and finally the controls themselves.

The following procedure, published by the Johnson Controls Company, is presented here to give you an idea of the kinds of instruction you can expect from manufacturers directed to field personnel.[3] It deals with the approach to checking out pneumatic controllers once it has been determined that the problem does not lie elsewhere.

(This analysis is based on a complaint resulting from improper temperatures.)

Check the set point of the controller for the correct setting. Check the output air pressure of the instrument controlling the air temperature in the complaint area. By examining the control diagram the proper output pressure for the actual temperature conditions existing at the thermostat can be determined. If, for example, an over heated condition occurs, the thermostat output pressure should be calling for the source of heating to be shut "off" and the

[3]Ibid.

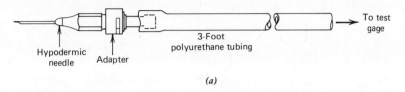

Hypodermic needle   Adapter   3-Foot polyurethane tubing   To test gage

(a)

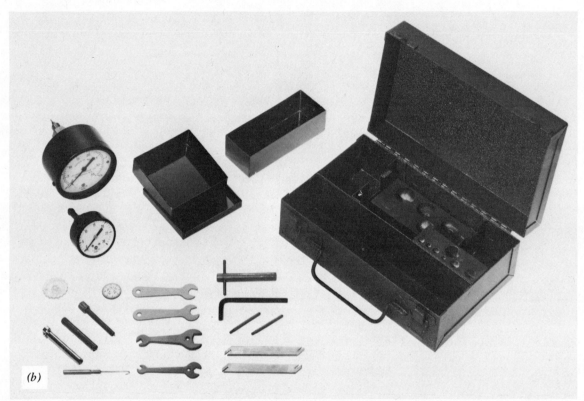

(b)

FIGURE 10–25   A hypodermic needle useful in reading pneumatic system pressures when connected to the test gage shown as part of the pneumatic service person's tool kit. The special pneumatic tools shown supplement a standard set of mechanics tools needed in everyday service work. (Courtesy of Johnson Controls Inc.)

source of cooling to be "on." If this is the case, it is not the thermostat causing the problem. If the thermostat output pressure is not as described above, the problem may be with the thermostat. Check the air supply pressure to the thermostat for correct pressure level and for any air leaks. If the above are OK, move the dial of the thermostat slowly from one extreme to the other to see if the output pressure will vary over the entire range of the supply pressure. If the pressure does vary, then it is probable that the thermostat is only out of adjustment and should be readjusted to the correct temperature setting. (See "Adjusting Instructions.")

If there is *no change* in output pressure by moving the dial, there is a malfunction in the thermostat. If only a partial change in pressure results, it can be the result of either:

1. malfunction in the thermostat
2. leak in the control pressure line
3. leak in the equipment being controlled

The repair of pneumatic components is best left to the manufacturer. Generally, depending on the device, a trade-in policy has been developed by the manufacturer enabling the customer to get a new or rebuilt part by turning in the defective part and paying a reduced price. In some cases detailed rebuild instructions are available for major items, but they are time-consuming to follow and the results are not guaranteed if done by a controls person not totally trained in the particular item. In the majority of cases, "Replace, do not repair" is a good philosophy.

DISCUSSION TOPICS

1. **Describe several methods of converting a change of temperature into a mechanical movement.**
2. **What is detent action and why is it used?**
3. **How are thermistors used in electronic control systems?**
4. **In what applications might an averaging sensor be superior to a standard sensor?**
5. **What kinds of systems require a changeover control? How may changeover be accomplished?**
6. **Describe how you would position a temperature control.**
7. **Why is 24-volt control circuitry superior to 230-volt control circuitry?**
8. **What is a deadband?**
9. **Explain the statement "Pneumatic systems are inherently proportional."**

The Formal
Plan

It is not enough that a method of controlling the environment of a building exist in the mind of the HVAC engineer. This method must be communicated to many people in order for it to take form in a completed structure. The owner must be informed so that he knows for what he is paying. The contractor must be informed so that he can develop a plan of installation and come up with a price. The installer must be informed so that he can install the system in accordance with the designer's idea. The service person must be informed so that he can determine whether or not the system is operating properly. The operating personnel must be informed so they can identify the equipment they must operate and ensure that it controls the building's environment within prescribed limits.

The formal plan is the committing to paper of the HVAC controls engineer's ideas. It is the written specification of the controls to be used and their sequence of operation. It is a series of directives that attempts to totally define the controls to be used, how they will be installed and how they will operate The specification is supplemented by design drawings that attempt to graphically portray the relationship between the HVAC system components and their controls.

In this chapter we investigate some commonly used methods of specifying controls and the types of control diagrams encountered in the field.

11.1 SPECIFICATIONS

Rarely does a building owner have the "know-how" to design an HVAC system complete with controls. A major exception is the large corporation that has an "in-house" facilities design capability. The owner of a small residential building may hire a contractor to design and install a system. A larger project owner may hire a consulting engineer to develop plans and specifications that are eventually submitted to a number of contractors for price bids. The larger the project, the more sophisticated the control documents. As the dollars involved increase, there is a greater need to guard against error and hence a greater need to completely describe the task to be accomplished. This is particularly true for work done for government agencies, where specifications can be thousands of pages long (even for small jobs, unfortunately).

What follows are some typical specification notes as they appeared in the specifications of a college building to be erected for a municipality. The notes on automatic controls covered 8 pages of the HVAC specification, which was 77 pages long. This, in turn, was part of the complete building specification, which, exclusive of change orders, was 840 pages in length. The HVAC specification was the longest of the individual crafts covered.

In this particular specification, four different bids were being requested by the owner: plumbing, electrical, HVAC and sprinkler, and general construction covering all other aspects of the building. On some jobs general contractors will be asked to bid the whole job. In turn, they will solicit bids from specialty companies, put together all the bids, and send one price to the owner. On larger jobs the breakdown into three or four major categories is becoming increasingly popular. A contract is awarded, presumably to the lowest responsible bidder, in each category. Note the term *responsible bidder*. In addition to offering the best price, the successful bidder also has to be able to prove that he is capable of doing the job.

The following has been extracted from the HVAC portion of a building specification. (The parenthetical comments are by the author.)

40. Automatic Controls

a. General
(1) Furnish and install, as hereinafter specified, a complete pneumatic temperature control system as manufactured by the "Honeywell Co.," "Johnson Control Co.," "Powers Regulator Co." (Note that three companies were stated as being the source of the controls. The contractor might have a good reason for using XYZ Co. controls, such as price or availability. If he goes to the engineer and owner and gets a written exception to that part of the spec he can substitute other controls. "Being Specced" is the goal of all controls sales engineers. Generally if they can get mentioned in an engineer's specification they have a good chance of making the sale.)
(2) Control systems shall be complete in every detail, including air compressors, air receivers, air dryer, automatic water valves, automatic steam valves, relays, dampers, motors, etc. which may be required to accomplish the end results as specified. (The key word here is "complete." An omission in the spec or plan doesn't relieve the controls contractor of his obligation to make the system work.

Spotting an omission or error early can save the contractor some money.)

(3)   The control system shall be installed by competent mechanics regularly employed by the manufacturer of the control equipment. All controls shall be the product of one manufacturer. (This requirement is that Honeywell, or Johnson, or Powers not only make the controls but *install* them as well. Widespread use of this requirement prevents small contractors from getting into the controls business.)

(4)   Complete control drawings shall be submitted for approval before any field installation is started. The drawings shall give a complete description of all control elements and shall show all schematic piping and wiring. In addition, the submittal shall include a complete description of operation. (By requiring the contractor to do this, the engineer attempts to ensure that any errors or omissions are found early. It is also a test of competence of the contractor.)

(5)   The Contractor shall familiarize himself with the Motor Schedule, and motor control sequence schematics on the contract drawings. The schematics should not be construed as being final detailed controls, but instead, as guide to the Electrical, Mechanical and Control Contractors. (This demonstrates that there is considerable leeway in putting together a control system. Two systems, installed from the same specs and plans, by two different contractors, can be substantially different.)

(6)   The pneumatic operators, shown on the motor control sequence schematics on the motor schedule contract drawing, are not necessarily the actual pneumatic operator to be used, but instead, as an indication as to how the pneumatic system is energized. (Actual part numbers may or may not appear on the engineering drawings. In this case they do not. Submittals by the contractor, however, must give such detail.)

(7)   The automatic control valves and control dampers furnished by this Contractor shall be installed by the HVAC contractor under the Control Contractor's supervision.

(8)   All interlocking wiring and temperature control wiring required to operate any automatic controls shall be furnished and installed by this contractor. (This is to reemphasize that the job is to be complete. Control wiring is not part of the Electrical Contract but rather the HVAC contract. Different electricians will do the work.)

(9)   The control system shall be guaranteed for a period of one year from the date of final acceptance against all defects in material and/or workmanship caused from normal use.

(This is standard and difficult. Typically, manufacturers warrant material for 1 year from the date of shipping. This equipment may be on the job site for 2 years before the job is finished and accepted. There is much negotiation between the HVAC contractor and the controls manufacturer when failures do occur in the warranty period.)

b. Air Compressors

(1) Furnish and install two high pressure, electric driven air compressors where indicated. Each compressor shall be of ample size so as to provide the necessary compressed air while operating not more than 33% of the time. (The necessary compressed air is determined by adding up the air requirement of all pneumatic components in the system.)

(2) Each compressor shall be provided complete with ASME labelled type air storage tank and shall be sized and furnished by the automatic control systems manufacturer.

(3) Compressor air storage tank units to have steel base ready for mounting on concrete base. Concrete base by section 0330. (This indicates that the concrete base for the compressor is to be prepared by the General Contractor.)

(4) The two compressors shall be arranged so that one acts as an automatic standby and will cut in whenever the other compressor fails. Sequence alternating switching shall be provided to index the lead compressor. (This insures that both compressors will have an equal amount of running time.)

(5) Furnish and install all necessary accessories for the air compressors including, but not limited to the following:
(a) Sequence alternating switch
(b) Safety relief valves
(c) Air filters and oil filters
(d) P.R.V. assemblies
(e) Belt guards
(f) Refrigerated after-cooler type air dryer
(g) Pressure gages
(Note the "but not limited to" phrase. Again, the system must be complete.)

(6) Each compressor shall have a combination magnetic starter, fused disconnect type, with a 120 volt control transformer. Furnish starter for installation by Electrical Contractor. (The electrical contractor is going to wire in the compressors, but the motor starter is to be provided by the controls contractor.)

(7) Compressors to be as manufactured by "Ingersol Rand Pump Co., Chicago Pump Co., Quincy Compressor Co."

c. Air Piping

(1) All air piping throughout the system shall be type "L" hard copper or soft shell seamless copper tubing run concealed where possible. Where piping is run exposed, only hard shell copper shall be used which shall be run horizontally and vertically plumb with reasonable pitch, provided with drip pockets, securely fastened at regular intervals, and run in a neat workmanlike manner.

Plastic tubing may be used within enclosed control panels, in fan coil enclosures, in Mechanical Equipment Rooms, in Mechanical shafts. Plastic pipe must conform with current General Service Administration Standards.

Tubing shall be protected by a polyvinylchloride (P.V.C.) sheath, electrical metallic tubing or sheetmetal raceway.

(2) Air pressure indicating gages shall be provided at all points throughout the system, where visual indication of air pressure is required for operating purposes.

d. Valves

(1) All valves shall be fully modulating unless otherwise specified. In addition, valves shall be quiet in operation, fail safe, be equipped with throttling plugs, and renewable composition discs, and be capable of operating at varying rates of speed to correspond with the exact dictates of the controller. Provisions shall be made for valves operating in sequence with operating ranges and starting points.

(2) All automatic valves shall be sized by control manufacturer to meet the heating or cooling requirements. (This is the function of the controls sales engineer in the field or the controls application engineer back at the manufacturer's home plant.)

e. Dampers

(1) All modulating dampers shall be of the multi-blade opposed action type.

(2) All two position dampers shall be of the multi-blade parallel action type.

(3) Dampers exposed to the weather, such as fresh air intake or exhaust air discharge, shall be constructed of No. 16 gage aluminum. All other dampers to be constructed of No. 16 gage steel.

(4) Damper blades shall be constructed with interlocking edges, lined with vinyl weather stripping. Maximum blade dimensions shall be nine inches.

(5) Damper linkage shall be rigged with a connecting rod which extends through the ductwork for connection to its

automatic operating device. Device shall be mounted on exterior of duct. (Damper operators will therefore be exposed for servicing.)

(6) The proper linkage shall be furnished to provide equal percentage or linear characteristics as required by the application.

(7) The control manufacturer shall submit charts used for sizing dampers and certification of damper leakage. (Damper leakage is a *major* cause of energy loss. The difference between low cost and expensive dampers is often the sealing technique used.)

(8) Pneumatic damper actuators shall be of the piston type. Pilot positioners shall be provided on each fresh air, return air and exhaust air dampers to assure that the dampers operate in unison. Pilots shall be provided for other sequencing applications as required.

(9) All dampers shall be provided in a mounting frame, with frame bolted to the ductwork. Frame shall be No. 16 gage aluminum or steel as required.

(10) Blades shall be mounted on rods set into nylon bearings. (To resist corrosion and insure smooth operation.)

(11) Dampers shall be furnished by the automatic control manufacturer.

(12) In general, in order to prevent drafts and protect equipment from the outside elements, each supply fan, return fan, exhaust fan, airconditioning unit and heating and ventilating unit shall be provided with an automatic damper on either the fresh air intake or discharge duct. Damper shall be normally closed and shall be arranged to open when its respective fan or unit motor is started and to close when the motor stops.

f. Duct, Pipe & Outdoor Sensing Thermostats

(1) All thermostats, except freezestats, firestats and all other clearly itemized thermostats, shall be pneumatic. These pneumatic controllers shall be fully proportioning, be provided with adjustable throttling ranges, and be of the relay non-bleed type unless specifically noted otherwise.

(2) All duct, pipe sensing, and outdoor sensing thermostats shall be provided with 2″ round air gages indicating main air pressure and branch pressure. Sensing elements shall be of the remote pneumatic transmission type. Capillary or transmitter sensing elements located in ducts shall have averaging sensing elements of sufficient length to cover a minimum of two thirds of the duct. Elements located in pipes shall be provided with wells. Pneumatic transmitters

shall be of the 100 degree range, 50 degree range or less as required and as approved. Transmitters of the 200 degree range are not acceptable except where temperature indicating or reset ranges exceed 100 degrees F.

g.   Room Thermostats

(1)   All room thermostats except unit heaters, cabinet heaters, and other clearly noted thermostats shall be pneumatic. These pneumatic controllers shall be fully proportioning, be provided with adjustable throttling ranges and be of the relay non-bleed type unless otherwise specifically noted.

(2)   Thermostats shall be provided with tamper proof devices.

(3)   Thermostat Covers—Covers for thermostats shall be neatly formed stamped steel. Thermostat covers shall be of heavy design to prevent damage. All necessary keys shall be furnished for removing the covers.

(4)   Thermostat Mounting—Each room thermostat shall be installed approximately 5 feet above the floor to center of thermostat on a neat cast iron or steel back, flush plaster and provided with lugs to secure cover and thermostat frame screws.

(5)   Thermostats for fan coil system shall be of the return air type and shall be mounted inside the custom built enclosures. An adjustable dial shall be mounted on the outside of enclosures.

h.   Thermometers

Thermometers shall be the pneumatic type. Each thermometer shall be a 3½″ dial type with a range such that the related controller setpoint falls approximately at the midpoint. Thermometers sensing elements shall match the related controller, i.e. averaging elements for averaging controllers, etc. All thermometers shall be flush mounted on panels and clearly labelled in an approved manner.

i.   Freezestats

(1)   All airconditioning systems shall be provided with a thermostat located on the upstream side of cooling coil or as indicated or scheduled on contract drawings and shall be arranged to stop supply fan when air temperature drops below 40°F.

(2)   Freezestats shall have twenty foot sensing elements responsive to the lowest temperature along the entire length. Elements shall be serpentined across the entire duct and located so they are not exposed to bypassed freezing air.

(3)   Where cooling coils are more than one bank in width, a thermostat shall be provided for each bank.

j. Smoke Detection Devices and Firestats

Firestats shall be provided for all exhaust air systems, and combination firestats and smoke detectors shall be provided for all return air systems. Devices to be provided, wired and connected by Electrical Contractor. HVAC Contractor shall determine location of the devices and shall install them. (The Fire Protection System is part of the Electrical Contract. The HVAC Contractor knows where they belong and so informs the Electrical Contractor.)

k. Pressurestats

All airconditioning supply and return air fans shall be provided with pressurestats which will shut down fans on an abnormal rise in pressure.

l. Master Control Panel

(1) Furnish and install an enclosed master control panel in the Mechanical Equipment Room.

(2) Panel shall be factory wired with a terminal strip, and/or piped, ready for field installation. Panel shall be provided with floor stands.

(3) All devices mounted on the panel or in panel shall have engraved bakelite nameplates, indicating system space title, and equipment tag number.

(4) Panel shall be the product of the automatic control manufacturer.

(5) Thermometers, switches, air gages, lights, etc, shall be flush mounted.

(6) Pneumatic controllers, pneumatic relays, electric controllers, electric relays, etc., shall be located inside the panel.

(7) Drawings and details for the master control panel shall be submitted for approval.

(8) Furnish and install on panel flush mounted indicating thermometers and controls as follows:

a. Outdoor dry bulb temperature
b. Outdoor wet bulb temperature
c. HTHW water return to North Academic Center
d. HTHW supply from North Academic Center
e. Chilled water supply from North Academic Center
f. Chilled water return from North Academic Center
g. Start-Stop and Run indication for all systems
h. Summer-Winter switches
i. Freezestat alarms for all HVAC systems
j. Space temperature indication for each HVAC system

(9) Furnish and install on panel a 12″ self-starting electric clock and fluorescent panel light.

FIGURE 11–1   The above photograph depicts a master control panel exterior. The larger dial faces show temperatures of air and water at remote locations as sensed and transmitted by pneumatic sensors. The smaller dials are air pressures for specific controllers. The photograph at right shows this same panel opened revealing gages, controllers, relays, and a time clock. This is the heart of the pneumatic control system.

(10) Panel shall be factory painted in color selected by the architect.

(11) All wiring and conduit from HVAC equipment shall be provided by the HVAC Contractor.

m. Sequence of Operation

(1) AC-1 Control

a. The supply fan and the return fan shall be electrically interlocked and shall be started manually from the central panel. During the unoccupied period a night thermostat shall cycle the supply and return fan to maintain its reduced night setting.

b. When the fans are started the EP switch shall be energized and the minimum outside air damper shall open. A receiver/controller located in the local panel shall sense the mixed air temperature through temperature transmitter TT1 and shall during the winter cycle modulate the variable outside, return, and spill air dampers to maintain its setting. During the summer cycle the variable dampers shall be in their normal position. (An EP switch is a solenoid valve, and a temperature transmitter is a pneumatic temperature sensor.)

c. Receiver/controller located in the local panel shall sense the supply fan discharge through temperature transmitter TT2 and shall control in sequence the cooling coil three-way valve and the preheat valve through a low limit thermostat located in the preheat coil discharge.

d. Prehcat coil inline circulating pump shall be energized whenever the mixed air temperature drops below 40 F.

e. Room thermostat shall modulate its reheat coil three-way valve to maintain its setting.

f. Low limit protection thermostat TE1 located in the preheat coil discharge shall stop the fan and energize an alarm at the central panel.

(2) Controls for Air Handling Systems AC-2, 3, 4, and 5

a. Same as AC-1

b. Same as AC-1

c. A low limit protection thermostat TE1 located in the heating coil discharge shall stop the fan and energize an alarm whenever the temperature falls below its setting.

d. Smoke detectors (furnished by the Electric Contractor) located in the return air intake and the supply air discharge. When energized shall stop the fan and close the outside spill air dampers and close the smoke dampers to inhibit the flow of smoke through the systems. When a smoke condition is detected an alarm shall be energized at the central fire alarm panel. A night room thermostat shall energize the supply and return air fans whenever the space condition falls below its setting.

(3) Fan Coil Control

A heating/cooling return air stat shall modulate a two-way valve on the combination coils to maintain its setting. A three speed fan control switch shall be supplied by the manufacturer.

(4) Primary Hot Water Control

a. A controller sensing the discharge temperature of the heat exchanger shall control a three-way valve on the high temperature hot water supplied from the central plant, to maintain its setting.

b. The primary hot water pump shall be started and stopped from the central control panel.

(5) Radiation and Unit Heater Control

A submaster thermostat with its bulb located in the hot water supply shall be reset from an outside air master and shall control a three-way valve from the primary hot water system to maintain its setting. The radiation inline circulating pump shall be energized whenever the outdoor air temperature falls below 60°F.

(6) Fan Coil Hot/Chilled Water Control

Summer/Winter switch shall index the three way switch over valves from hot water to chilled water. During the summer condition the chilled water shall be supplied from the central plant and shall not be controlled. During the winter condition a controller located in the three way mixing valve discharge shall be reset by the master outside air controller and shall modulate the three way mixing valve to maintain its setting. A differential pressure controller shall control the differential pressure between the hot/chilled water supply and the hot/chilled water return by modulating a normally open differential pressure valve.

(7) Exhaust Fan Control

The fans shall be started from the central control panel. When the fan starts, the exhaust dampers shall be opened; when the fan stops, the exhaust dampers shall be closed.

(8) Future Automation

All controls shall be capable of interfacing with a future central automation system, to indicate and to reset temperatures, to start/stop motors, and to indicate alarm conditions.

11.2 CONTROL DIAGRAMS

The drawings of the HVAC system, referred to as "mechanicals," are intended to expand upon the words in the specification. In Figure 11–2, for instance, we see the drawing of an air handling system AC-1. Before looking at the control details, we consider the entire drawing. Note the return and supply air fans, which are to be electrically interlocked; the presence of a spill air damper, a return air damper, and maximum and minimum fresh air intake (FAI) dampers. The mixed return and fresh air passes through the preheat coil (PHC), the cooling coil (CC), and through the reheat coil (RHC). Only one RHC is drawn, but it is actually typical of eight such arrangements as indicated by the note "typ for 8" written below the designation "room thermostat." It is always a good idea to get an overview of the system before studying the details.

When called in to start-up or service a system, the controls person would be well advised to walk around the building first, under the guidance of the building operating personnel to get an idea of the physical layout before delving into the drawings. While walking around, questions should be asked. Contrary to what many think, there is nothing wrong with asking questions. After examining the drawings, perhaps another walk through the building would be useful. The main idea of this orientation procedure is to get comfortable with the building system.

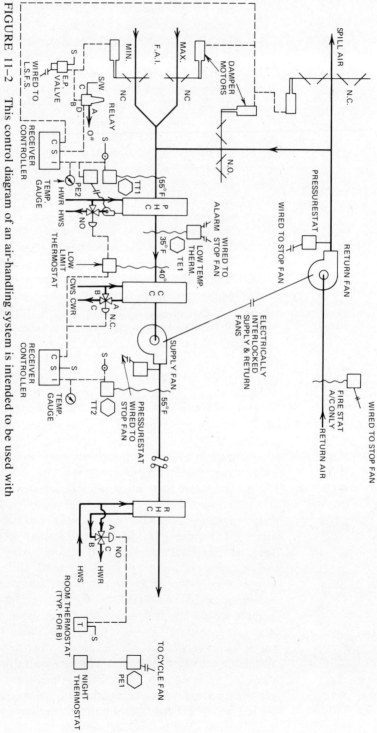

FIGURE 11-2   This control diagram of an air-handling system is intended to be used with the sequence of operation to define how the system works. Note that the controls manufacturer and part numbers are not indicated on this drawing.

It would be ideal if every engineer and draftsman in the world used the same symbols and letter abbreviations to signify particular items in the control drawing. Although efforts have been made from time to time to develop industrywide drafting standards, such standards, while helpful, will not be followed rigidly by everyone. This, and also the fact that the people originating the drawings are human and therefore subject to omissions and errors, can make comprehension of drawings a challenge. The more experience we have, the faster we are able to understand what the draftsman is trying to communicate. The more complete our understanding of air conditioning principles, control principles, and available control hardware, the better able we are to identify errors and fill in information that has not been provided. Incidentally, if there is a question about a drawing, where a wrong assumption on our part could have far-reaching results, we can call the people who made the drawing and get clarification. This may at times be a bit tedious, but corrections made over the telephone are much less costly than those made on the job.

Figure 11–3 includes a symbol list. Presumably, any symbol or abbreviation used on the drawings can be found on this list. Also, some symbols used on the drawings may not be listed. As an example, FAI is used to describe *f*resh *a*ir *i*ntake, yet the symbols list does not contain this abbreviation. It does, however, have OAI, or *o*utside *a*ir *i*ntake, which is the same thing. If we are flexible and use our imagination and knowledge, the parts of the drawing will usually all fall into place.

11.2.1 Air Handler AC-1

Let us go through the specification for the Sequence of Operation of Air handler AC-1 [paragraph m(1) of the specification] and see how the diagrams reflect the written word.

The requirement for interlock between supply and return fans can be seen by reading the note next to the line drawn between the two fans. Figure 11–4 is a typical ac unit wiring diagram, which indicates how the supply fan is controlled. In Chapter 9 we went into the details of how a motor starter works. In Figure 11–4 we should recognize the "start-stop" button arrangement as well as the three overloads (OL), one for each of the three-phase wires, and the starter coil labeled AC.

Between the start-stop buttons and the motor starter coil are two paths for the electricity to follow. Which path is followed is determined by the selector switch labeled A (for automatic), O (for off), and H (for hand). With the switch in the A position the path of electricity is through a freezestat and pressurestat. Let us assume this to be the case. Now we look at the four switches between the pushbuttons and the motor starter coil and relate them to the specification.

SYMBOL LIST

Symbol	Description	Symbol	Description
	LOW PRESSURE STEAM		PRESSURE GAUGE CONNECTION
	LOW PRESSURE STEAM CONDENSATE RETURN		AUTOMATIC AIR VENT
	CONDENSATE OR VACUUM PUMP DISCHARGE		MANUAL AIR VENT
—CHWS—	CHILLED WATER SUPPLY		CLOSED BUCKET TRAP
—CHWR—	CHILLED WATER RETURN	F&T	FLOAT AND THERMOSTATIC TRAP
—CHS—	DUAL TEMPERATURE WATER SUPPLY		FLANGED UNION
—CHR—	DUAL TEMPERATURE WATER RETURN		SCREWED UNION
—HWS—	HOT WATER SUPPLY		GATE VALVE
—HWR—	HOT WATER RETURN		GLOBE VALVE
—ED—	EQUIPMENT DRAIN		BALANCING AND SHUTOFF VALVE
	COLD WATER OR CITY WATER MAKEUP		LUBRICATED PLUG VALVE
—CA—	COMPRESSED AIR		CHECK VALVE
—VENT—	VENT		STRAINER
	DIRECTION OF FLOW		STRAINER WITH BLOWOFF VALVE
	ANCHOR		SPRING-LOADED SILENT CHECK VALVE
	EXPANSION JOINT		HOSE END DRAIN VALVE
—FH—	FLEXIBLE HOSE CONNECTION		PRESSURE REDUCING VALVE
—PDN—	PITCH PIPE DOWN		PRESSURE AND/OR TEMPERATURE RELIEF VALVE
	THERMOMETER		TWO-WAY CONTROL VALVE
	THERMOMETER WELL		THREE-WAY CONTROL VALVE
	WATER PRESSURE GAUGE	T	ROOM THERMOSTAT
		T$_N$	NIGHT THERMOSTAT
		T○	OUTDOOR THERMOSTAT

(H)	HUMIDISTAT		BR	BOTTOM REGISTER
(F)	FIRE STAT		CD	CEILING DIFFUSER
(S)	SMOKE DETECTOR		CG	CEILING GRILLE
[SW]	SUMMER/WINTER SWITCH		CR	CEILING REGISTER
[PC]	PROGRAM CLOCK		TG	TOP GRILLE
↑	SUPPLY AIR		TR	TOP REGISTER
↓	RETURN AIR		TG	TRANSFER GRILLE
	SUPPLY OR OUTSIDE AIR DUCT		N	NECK
⊠	RETURN OR EXHAUST AIR DUCT		OAI	OUTSIDE AIR INTAKE
R→	INCLINED DUCT RISE		EXH	EXHAUST AIR
D→	INCLINED DUCT DROP		RA	RETURN AIR
VD	OPPOSED BLADE VOLUME DAMPER		SA	SUPPLY AIR
	DUCT TURNING VANES — DUCT DIMENSIONS SHOWN ARE INSIDE CLEAR		SD	SPLITTER DAMPER
	FLEXIBLE CONNECTION		FC	FLEXIBLE CONNECTION
	ACCOUSTICAL LINING		HOA SW	HAND OFF AUTOMATIC SWITCH
	ADJUSTABLE DEFLECTOR VANES		LP	LOW PRESSURE
SD	SPLITTER DAMPER		NC	NORMALLY CLOSED
M	MOTORIZED DAMPER		NO	NORMALLY OPEN
⊠	BLANK OFF DIFFUSER		NTS	NOT TO SCALE
AD	ACCESS DOOR		CFM	CUBIC FEET PER MINUTE
FD	FIRE DAMPER		₵	CENTERLINE
FD&AD	FIRE DAMPER AND ACCESS DOOR		(SD)	SMOKE DAMPER WITH DETECTOR
BG	BOTTOM GRILLE		LSFS	LOAD SIDE FAN STARTER

FIGURE 11–3 The symbol list provides the key to understanding the intent of the drafter who made the drawing. In some cases, such as this, a standard symbol list is used containing a number of symbols not appearing on the drawing. Unfortunately, some symbols that do appear on the drawing are not contained in the list.

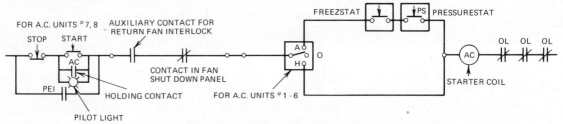

FIGURE 11–4   A typical air conditioning unit wiring diagram. In this case the air conditioning unit is actually only an air handler, that is, supply fan and associated controls.

The first switch to the right of the pushbuttons is labeled "auxiliary contact for return fan interlock." Figure 11–5 is the typical wiring diagram for return air fans and is very similar to the arrangement for three-phase motors we have seen before. When the start button is pushed, it energizes the RF coil, which closes the switches to the return fan motor (not shown in this diagram) and *also* closes the *auxiliary contact* shown in Figure 11–4. If this auxiliary switch is not closed, the supply fan cannot run. Said another way, if the return fan is not energized first, the supply fan cannot be energized. This is the meaning of the term *interlock*. The requirement of paragraph m(1)a for interlock is met by using this auxiliary contact.

The next switch to the right is labeled "Contact in Fan Shut Down Panel." It is a normally closed switch, but its description is not totally clear. There is a requirement in paragraph j that each return air system have a firestat—one is shown in the AC-1 diagram. One would surmise, then, that this curiously labeled contact is built into the firestat, which is "wired to stop fan." If the firestat sensed an excessive temperature, say in excess of 125°F, the contact would open, the starter coil would be deenergized, and the fan would shut down.

Skipping over the "A-O-H" switch, we see the freezestat. Paragraph m(1)f calls for low temperature protection provided by TE1, which stops the fan. Note that TE1 on the drawing (Figure 11–2) has a cutout temperature of 35°F showing, although in the specification 40°F is indicated. Since we are being given a choice, the wise selection would be 40°F as the setting. This is a safety device; 40°F will be safer then 35°F.

The fourth switch is labeled PS for pressurestat and fulfills the requirement of paragraph K. In the event that the duct pressure at the fan outlet becomes excessive, this switch, sensing the high pressure, will shut down the fan.

The requirement of paragraph m(1)a for night operation is met by using a pneumatic-electric relay labeled PE-1. In Figure 11–2 it is

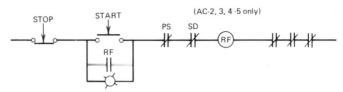

FIGURE 11–5   A typical wiring diagram for return fan. The expression "typical" means that although there may be several such fans in the building, only one wiring diagram is presented because each fan has the same wiring configuration.

shown connected to the night thermostat. In the wiring diagram (Figure 11–4) the PE switch bypasses the manual push buttons and provides automatic operation only at night. The supply fan cycles on and off during night hours in response to the night thermostat, which during the heating season is set several degrees lower than the day setting and during the cooling season is set several degrees higher than the day setting. The expression "night setback" is used to describe this feature, an energy conservation device very commonly employed on newer systems.

To this point we have considered only fan control. Now we move on to the controlling temperatures in the building.

The first sentence of paragraph M(1)b calls for minimum outside air when the supply fan is started. In Figure 11–2 this is accomplished with an EP valve "wired to LSFS" (load side fan starter). When the fan motor starter energizes, the fan motor simultaneously energizes the EP valve allowing full pneumatic supply pressure to the minimum FAI damper motor causing that damper to open. The balance of that paragraph pertains to maintaining mixed air temperature before the preheat coil. The receiver-controller is activated by a signal from the changeover relay. This relay receives a signal for summer or winter operation at port C (note S/W for summer/winter). In winter the signal to the S port of the receiver-controller causes that device to modulate the FAI maximum damper motor, the spill air damper motor, and the return damper motor to maintain 55°F at the TT1 sensor. This sensor is sending a pneumatic signal to port I of the receiver-controller, which causes the appropriate control signal to be sent from the C port. In summer the receiver controller sends a zero control signal to the three dampers, which assume their normal position, NC for FAI maximum, and spill air and NO for return air.

While looking at the pneumatic system around TT1 and the receiver controller, note PE2. This fulfills the requirement of paragraph m(1)d. When the temperature falls to 40°F, the pneumatic signal from TT1 causes PE2 (pneumatic-electric relay) to energize the motor driving the preheat coil pump.

The control of the PHC and CC is described in paragraph m(1)c. Temperature transmitter TT2 sends a signal to the receiver-controller which in turn sends a signal to the control valves on both coils. These valves are designed to act sequentially; that is, a gradually increasing temperature at the TT2 sensing point will cause a pneumatic signal that gradually closes the PHC control valve, passes through a deadband, and then gradually opens the CC control valve. This sequencing was specifically described in a previous chapter. The low limit thermostat insures that in a low temperature situation, even if TT2 does not call for heating, the PHC control valve will open. In essence the low limit thermostat dumps any positive pressure from the receiver-controller, creating a low pressure signal, which opens the PHC valve.

The final part of the AC-1 control system is the reheat coil control (RHC), of which there are eight. A simple pneumatic thermostat sends a signal to the RHC control valve to add that bit of heat necessary to maintain the temperature at the set point in the particular zone.

This diagram is a directive to the contractor indicating how the system should work. The contractor must select specific hardware made by Johnson, Honeywell, or Powers [refer to specification paragraph a(1)] to do the job as described. The contractor in turn makes a submittal drawing very similar to the system drawing we have been studying but which includes the manufacturer's part numbers and shows actual pneumatic line connections and specifications. The submittals, when accepted and approved by the owner or a representative (usually the engineer on the job) become the installation drawings for the installers in the field.

11.2.2 Air Handlers 2, 3, 4, and 5

The control diagram for air handlers AC-2, 3, 4, 5 is similar to that for AC-1 except that smoke detection and smoke damper control has been added. In Figure 11–6 note that the smoke detectors are installed before the return fan and after the supply fan. The note "wired to CFAP," next to the smoke detector, might be confusing, since no such abbreviation appears in the Symbols List. Reading paragraph m(2)d, however, leads us to surmise that CFAP is the *c*entral *f*ire *a*larm *p*anel.

If smoke is detected in either the return or supply air ducts, *both* fans must shut down. When that occurs, the EP valve, which is wired to the load side of the fan starter, will close and the pneumatic supply pressure will drop to zero. All dampers, including the smoke dampers, will go to their normal position. The smoke dampers, being labeled N.C., will close. To the right of the summer/winter relay, note "to points marked EP"; this indicates that pneumatic system pressure is to be carried to the point marked

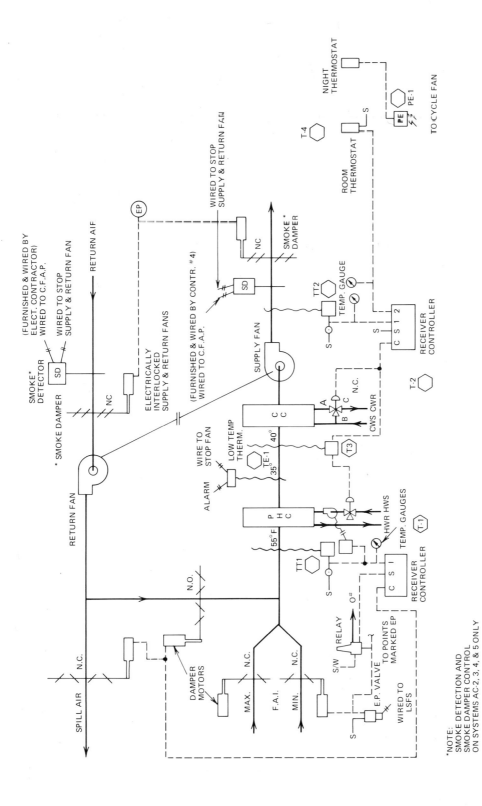

FIGURE 11-6 Compare the air handler system shown here with air handler system 1. The major difference is the smoke detection equipment and smoke dampers.

EP, which is connected to the smoke dampers. This is just a drafting technique used to eliminate a couple of lines from the drawing. In the field we would expect to see a copper line carrying pneumatic supply air at 15 psig to that point.

11.2.3 Fan Coil Control

The control diagram for the fan coil is shown in Figure 11–7. The specification in paragraph m(6) calls for a return air thermostat which might suggest that it be built into the fan coil unit; however, the diagram seems to indicate a remote wall-mounted pneumatic thermostat. Usually, consulting that part of the specification describing the fan coil will provide the appropriate information. If not, call the engineer and have the problem resolved.

Note that in this unit a single coil is used to carry both hot and chilled water depending on the season. The two-way modulating valve is normally open so that as previously described, in the event of an air pressure failure, heat would still be available in winter, although uncontrolled, to prevent a freezeup.

11.2.4 Water Systems

In Figure 11–8 the hot water control diagram shows how the requirements of paragraphs m(4), (5), and (6) will be met.

The source of heat is a central plant, external to the building, generating high temperature hot water (HTHW) at 375°F and distributing it to a number of buildings on the college campus. This is commonly referred to as "district heating." The HTHW flows through a heat exchanger, where it heats the water to be used to heat the building. The HTHW flow is regulated by a three-way valve controlled by a controller sensing heating system, hot water

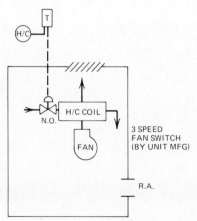

FIGURE 11–7   Fan coil control diagram. The internal wiring of the fan coil unit is done by the manufacturer so that it is not shown on the mechanical drawings of the HVAC system.

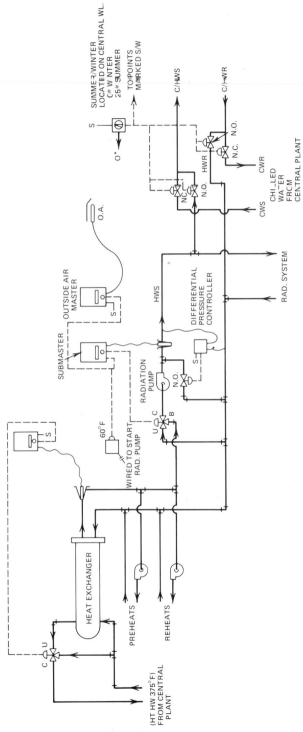

FIGURE 11-8   The source of heat for this building is high temperature hot water (HTHW) generated at a remote site and piped to the heat exchanger in the building. In some applications steam is piped in and passed through a similar heat exchanger, referred to as a converter.

supply temperature. This hot water is pumped to the preheat and reheat coils by manually started pumps.

Hot water from the heat exchanger is also available to the control system feeding the perimeter radiation, unit heaters, and fan coils. The temperature of this water, labeled HWS, is varied, depending on the outdoor temperature. An outside master thermostat resets a submaster and also energizes an in-line radiation pump when the outdoor temperature falls below 60°F. The submaster controls a three-way valve that mixes hot water from the heat exchanger and return water from the perimeter system to achieve the desired water temperature.

Two-way changeover valves are used to send chilled water to the cooling coil (CC) of the air handlers and to the fan coil units. They are two-position pneumatic valves that are manually set. A zero signal in winter closes the chilled water valves (NC) and opens the hot water valves (NO). A 25 psig signal in summer opens the chilled water valves and closes the hot water valves.

In a zoned system with modulating valves, if the heating demand throughout the building decreases, the valves will close off. Pressure at the discharge of the pump increases as the restriction to the flow increases. To prevent this, a differential pressure controller is used. As the modulating valves shut down more and more water is bypassed around the system to the return and the pump discharge pressure remains relatively constant.

11.2.5 Exhaust Fans

In Figure 11–9 the requirement of paragraph m(7) is depicted. The exhaust fans are manually started with a pushbutton electrical motor starter from the main control panel. Wired to the load side of each motor starter is an EP valve. When the motor is started, the valve opens allowing air pressure to act on the exhaust fan damper actuator and then open the damper. When the motor is shut down, the pressure signal drops to zero and the damper closes.

11.3 SCHEDULES

In addition to control drawings and specifications, schedules of equipment are included as part of the HVAC system documentation. These schedules list the equipment used and give technical information of value to contractors, salesmen, installers, and service people. Schedules of air handlers, unit heaters, fan coils, fan motors, pump motors, and other equipment are commonly presented in a format similar to that shown in Figure 11–10.

Figure 11–10 contains a schedule of motor starters. The information here is of interest to the electrician because the installation will be done by the electrical contractor. It is also of interest to the HVAC controls person, since the starters must be energized to

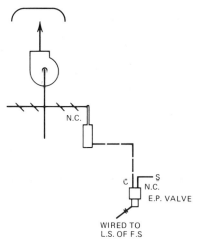

FIGURE 11-9   A good example of electrical and pneumatic controls being used together. The fan motor is energized manually by a pushbutton motor starter and the dampers are opened by a pneumatic actuator controlled by an E.P. (solenoid) valve.

operate the HVAC system. Let us consider two typical starters and see how the information is presented.

Motor starter M3 will have a nameplate inscribed with the word "stage" on its first line and "AC#4" on its second line. This indicates that air handler number four services the stage area. The motor controlled by M3 is in AC#4 located in the roof mechanical equipment room (MER). The motor has a 15-horsepower rating and draws 43 amperes full load (FLA). The starter itself is categorized as a NEMA Size 3 magnetic starter with one pushbutton and two normally open auxiliary switches and two normally closed auxiliary switches. The starter is located on panel MCC-1 and is to be wired in accordance with the "typical A.C. unit wiring diagram." The electrical disconnect switch in line with the starter is to be of a 100-ampere size and contain 70-ampere fuses. The wires to the starter are to be three #6 AWG (American Wire Gage) conductors contained in a 1-inch-diameter conduit.

Motor starter M14 controls a ceiling exhaust fan motor labeled E-2. (The prefix AC stands for the supply fan motor, "R" for a return fan motor, and "E" for an exhaust fan motor.) The exhaust fan motor is also located in the rooftop mechanical equipment room. The motor has a 1-horsepower rating and draws 4.0 amperes. The starter is a NEMA size 00 (two nought) magnetic type with one pushbutton and two normally closed and two normally open auxiliary contacts. (Auxiliary contacts, it may be recalled, have a lower current-carrying capacity than the main starter contacts and are used in control circuitry such as the supply/return

No.	Motor Starter-Motor Center Name Plate Inscription		Motor				Starter				
	First Line / Service	Second Line / Equipment	Location		H.F.	F.L.A.	NEMA Size	Type	Button Switch Pilot	Auxiliary Contacts	
			Floor	Location						NO	NC
M1	Music hall	AC #2	Roof	M.E.R.	20	55	3	Mag	1	2	2
M2	Theater	AC #3			15	43	3		1	2	2
M3	Stage	AC #4			15	43	3		1	2	2
M4	Lobby	AC #5			15	43	3		1	2	2
M5	Shops	AC #6			15	43	3		1	2	2
M6	Toil, Exh.	E-1			¾	3.1			1	2	2
M7	AC #2	R-2			7½	24.0	1		1	2	2
M8	AC #3	R-3			3	9.5	0		1	2	2
M9	AC #4	R-4			3	9.5	0		1	2	2
M10	AC #5	R-5			3	9.5	0		1	2	2
M11	AC #6	R-6			3	9.5	0		1	2	2
M12	Paint shops	E-3			1½	5.3	00		1	2	2
M13	P.H. M.E.R.	E-9			1	4.0	00		1	2	2
M14	Cel. Exh.	E-2			1	4.0	00		1	2	2
M15	Stage Exh.	E-5	↓	↓	1½	5.7	00	↓	1	2	2

| Remote Station | | Motor Starter Location | Control Diagram | | Switch | | Feeder | Conduit |
Location and Type	Button Switch Pilot		For A.C. Unit	For Fan	Switch Size	Fuse Size		
CELLAR MECH. RM. START-STOP PNL.	1	MCC-1	√		100	100	3#4	1¼″
CELLAR MECH. RM. START-STOP PNL.	1	MCC-1	√		100	70	3#6	1″
CELLAR MECH. RM. START STOP PNL.	1	MCC-1	√		100	70	3#6	1″
CELLAR MECH. RM. START-STOP PNL.	1	MCC-1	√		100	70	3#6	1″
CELLAR MECH. RM. START-STOP PNL.	1	MCC-1	√		100	70	3#6	1″
CELLAR MECH. RM. START-STOP PNL.	1	MCC-1		√	30	15	3#12	¾″
CELLAR MECH. RM. START-STOP PNL.	1	MCC-1		√	60	40	3#8	¾″
CELLAR MECH. RM. START-STOP PNL.	1	MCC-1		√	30	15	3#12	¾″
CELLAR MECH. RM. START-STOP PNL.	1	MCC-1		√	30	15	3#12	¾″
CELLAR MECH. RM. START-STOP PNL.	1	MCC-1		√	30	15	3#12	¾″
CELLAR MECH. RM. START-STOP PNL.	1	MCC-1		√	30	15	3#12	¾″
CELLAR MECH. RM. START-STOP PNL.	1	MCC-1		√	30	15	3#12	¾″
CELLAR MECH. RM. START-STOP PNL.	1	MCC-1		√	30	15	3#12	¾″
CELLAR MECH. RM. START-STOP PNL.	1	MCC-1		√	30	15	3#12	¾″
CELLAR MECH. RM. START-STOP PNL.	1	MCC-1		√	30	15	3#12	¾″

FIGURE 11–10   Equipment schedules are a convenient way of identifying motors, air conditioning units, grilles and registers, and most other "store bought" parts to be used on a job. They are particularly valuable in locating the components in the system.

interlock). The starter is located on panel MCC-1 and is to be wired in accordance with the "typical fan wiring diagram." The disconnect switch used with this starter is to have a 30-ampere rating with 15-ampere fuses being used. The wires will be three #12 AWG conductors encased in a ¾-inch conduit.

DISCUSSION TOPICS

1. What is the value of a written sequence of operation?
2. What is the purpose of the control specification?
3. What is meant by the term *lowest responsible bidder?*
4. What are schedules of materials?
5. What are interlocking controls?
6. Paragraph 40(e) of the sample control specification deals with dampers and damper leakage. What are some problems associated with damper leakage?
7. What is the purpose of the thermostat TE1 described in paragraph 40(m)2c of the sample control specification?
8. What is district heating?

Chapter Twelve Building Operations Supervisory Systems

In a simple single family home, building operations are quite basic. An uncomfortable temperature causes the homeowner to adjust the thermostat. If the owner smells smoke, he or she will search for the cause and upon detecting a fire will call the fire department. If an intruder tries to gain entry, the homeowner, upon hearing these efforts, will call the police. A breakdown in the heating system results in discomfort followed by a call for service. If the lights are burning in an unoccupied room, energy is being used needlessly, and upon discovering the burning lights, the homeowner will inevitably shut them off.

There are a number of devices on the market that attempt to relieve the homeowner of these tasks. Smoke and fire alarms with a telephone dialing feature alert the fire department. Thermostats with a night setback feature, light switches with a built-in timer, and burglar alarm systems tied in to the police department are examples of such devices. Although there is a market for such items, they are generally considered "luxury" items and are not found in the typical residential situation.

As the application increases in sophistication and size from the single family residence to small commercial to large commercial and institutional complexes, the degree of automation in controlling the above functions, and many others besides, increases drastically. In recent years automated building systems operation has grown at an increasing rate and has been given a further impetus by the growing concern for energy conservation. In this chapter we discuss approaches to automated building operation supervisory systems.

12.1 SUPERVISORY SYSTEMS

Until rather recently most HVAC systems were monitored and adjusted manually. In smaller installations a janitor or maintenance man, generally an all-purpose handyman, had responsibility for the HVAC equipment. In larger systems round-the-clock "watch engineers" did just that; they watched the HVAC equipment. It is not unusual to find systems, originally designed with automatic controls, that have had the controls disconnected and are now operated manually by the watch engineers. The reliability of the early systems was questionable. Whether done automatically or manually, a number of operations must be accomplished in dealing with building systems.

Equipment must be started and stopped. Its operation must be monitored to see that it is doing the job desired and is functioning properly. Abnormally high or low temperatures, pressures, current draw, and other operating characteristics must be observed. Recording of system variables over a period of time is useful in detecting operating trends that might point to the need for repair. Records should be maintained indicating when and what maintenance is required and has been done. Proper control point values for particular systems under particular climatic conditions must be determined. Adjustments to achieve these control points must be made. Determination of required maintenance, either on a scheduled or "as needed" basis must be made. The performance of the equipment should be monitored to insure that it is at peak efficiency.

In a small building this is relatively simple, but obviously as the complexity of the installation increases, the effort, hence dollars, to adequately supervise the system becomes significant. The result is that either large amounts of money are spent to do it right, or it just is not done right with resulting premature equipment failure, downtime, and uncomfortable conditions in the building. The process of automating a building operations system is nothing more than taking jobs done manually and having them done by machine.

As in most automation changeovers some jobs are lost and others are gained. The sociological impact of these systems is beyond the scope of this book, yet should not be overlooked when changeover to an automated system is being considered. The technological impact, however, is most important and quite positive.

In justifying the change from a manual to an automated system, there are usually five main considerations:

1. The new system should require less manpower and therefore result in labor cost savings.
2. It should result in the conservation of energy and therefore produce lower system operating costs.

3. It should result in longer equipment life, hence stretch out replacement time with a lower annual equipment cost.
4. It should spot malfunctions early enough to prevent major equipment failure.
5. It should provide data upon which sound management decisions can be based.

It can accomplish some or all of these objectives by automatically doing the following: reducing the running time of the equipment, using outside air for cooling in an economizer cycle application, optimizing control points to provide comfort based on outside conditions, reducing loads (load shedding) to minimize the energy cost penalty during peak power consumption periods, tying into fire and security alarm systems, providing communications in the event of emergency, controlling lighting levels, controlling elevator operation based on usage, monitoring system variables and operating characteristics, and managing the maintenance program. The manner in which these objectives can be accomplished is the subject of the next sections.

12.2 CENTRALIZED SYSTEMS

Centralized automated systems can be considered to be made up of input devices, transmission links, central equipment, and output devices.

12.2.1 Input Devices

These are the sensors that monitor the condition of a variable. A thermistor is one such device in which the electrical resistance changes with the temperature. The input devices are remotely located and generate signals analyzed by the central equipment. If the signal generated by the device is an infinitely variable one, such as a changing temperature translated into changing resistance, the device is called an "analog" sensor. If the signal generated has discrete values, such as a temperature of 100.5°F, it is said to be a "digital" sensor. Actually, in order to make sense to us humans, the information we must have will usually be in digital form; however, some computers use analog *input* information, which eventually is presented to us in a digital format.

In attempting to analyze an air conditioning problem, the service person may find it useful to measure air or water temperature at a number of points. As mentioned earlier, larger systems have thermometers located at strategic points in the system. The service person must locate these thermometers, must reach them and read them (sometimes a difficult task in a poorly lighted area), and if a thermometer has not been installed where desired, must attach a service instrument and take a reading. In practice, as few readings

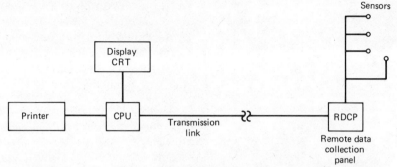

FIGURE 12–1   Components of a typical automated system. The input devices (sensors and remote data collection panel), transmission link, central processing unit, and output devices (printer and/or cathode ray tube display).

as possible are taken because of the time and effort involved in gathering all the desired readings. Imagine how useful an information system could be if at the touch of a button remote temperature and pressure readings would appear on a screen.

The term *computerized* is used popularly to describe systems that are primarily "information gathering" in nature. Although there may be some limited computational function built into these so-called computers, they are primarily information management systems.

The input devices must be compatible with the central equipment in these systems. They must be able to put the information in a form the central equipment can understand. A watch engineer can call the supervisor on a telephone and say "The condenser water temperature is 125 degrees Fahrenheit" and the supervisor will understand him. The input device must translate a temperature reading into an analog or digital signal or into an electronic coded signal that the central equipment can understand.

It is not inconceivable that hundreds or thousands of points within a building are being monitored. One major factor in the cost of such automated supervisory systems is the number of points to be monitored. These points may be measuring temperature, pressure, and humidity; pump motor current draw; or the status of a smoke detector. They may also be determining whether or not a motor is running or whether a particular door is open or closed. With a bit of imagination one can probably define several thousand bits of information that would be helpful to know about a building of a fairly modest size.

Several buildings can be linked together with central equipment at one point. In fact, a service offered by some companies has buildings dozens of miles apart linked to a central facility. Even though these buildings have different owners at widely scattered

FIGURE 12–2   This is an electronic humidity sensor used with an automated building system. The signal generated by this device goes to the remote data collecting panel, where it is coded and sent on to the central equipment. (Courtesy of Robertshaw Controls Company.)

locations, the information goes to one central point where it is monitored and analyzed. Each building owner is billed separately for this service depending on how many input points are connected. Problems showing up at the central station are communicated by telephone to the appropriate building operating personnel for action. It is common in such situations to have remote data collection panels at each site to collect information from the input sensors, organize it, code it, perhaps store it for later use, and generally to prepare it for transmission to the central equipment. This panel, although separate from the various sensors, is considered to be an input device as far as the central equipment is concerned.

12.2.2 Transmission Links

The information collected at a remote point must be transmitted to the central equipment. The electronic signals are quite sensitive to outside electrical influences and therefore the transmission lines are considered with great care. In small systems with a single building or within a tight complex of buildings, it is not unusual to

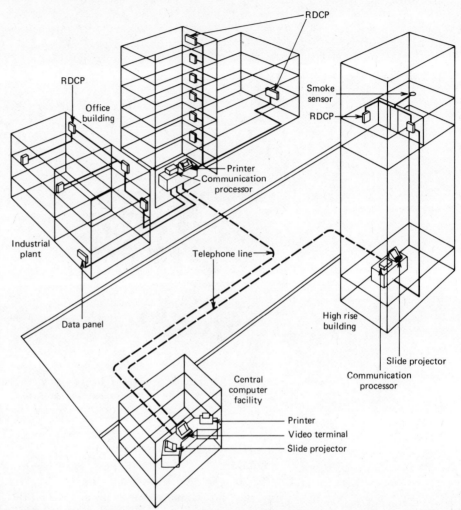

FIGURE 12–3 A regional network configuration (several buildings linked together in a network). The buildings can be in a complex close together or separated by many miles. Each building may have a small computer and also be tied in to a larger capacity CPU at the central facility. This is another example of the flexibility of such an arrangement.

have wire installed specifically to handle the electronic traffic of the HVAC supervisory system. Early systems employed multiconductor cables with dozens or hundreds of conductors depending on the complexity of the system. Newer systems use two conductor wires referred to as "twisted pairs" to carry coded signals. The high cost of labor in laying long runs of expensive cable has been overcome by increasing the complexity of the input and central equipment at relatively little cost by means of low cost microprocessor chips.

FIGURE 12–4   This remote data collecting panel not only collects and organizes the signals from remote sensors but also provides a speaker system to enable a service person at the remote location to speak to the central console operator. Such communication is valuable in tracking down system problems. (Courtesy of Robertshaw Controls Company.)

It is important to note that a "system" *includes* the transmission links and the manufacturer's recommendations as to wire type and installation procedures that must be followed.

In systems covering wide areas, communication is accomplished by telephone lines. In some computer systems a remote terminal user dials a telephone and lays the receiver on a special cradle to achieve contact with the central computer. In HVAC control systems the telephone company actually runs the phone lines directly to the remote data collection panels and the central equipment. Input signals are transformed into signals that will carry well over the telephone wires by a device called a "modem." This modem may be a component part of a remote data collection panel. Another modem at the central equipment location will take the telephone signals and transform them back into electronic signals compatible with the central equipment. These telephone lines are, for obvious reasons, called "voice grade" lines and have the ability to carry a high rate of electronic traffic. Earlier systems used "telegraph grade" circuits with a significantly lower transmission rate. While much higher rates of information transmission can be achieved with specially installed (referred to as "dedicated") twisted pairs, the leased telephone lines are the most economical and the most widely used today.

12.2.3 Central Equipment

The central processor is a collection of equipment that receives input information, organizes it, and generates output information. Although a computer may be included in the central processor, the main function of the CPU is to receive, transmit, and present information.

FIGURE 12–5   An interior view of a central electronics cabinet. Shown, from top to bottom, are an audio amplifier, dual disk memory, computer, communication interface cards, modems, and power supply. (Courtesy of Robertshaw Controls Company.)

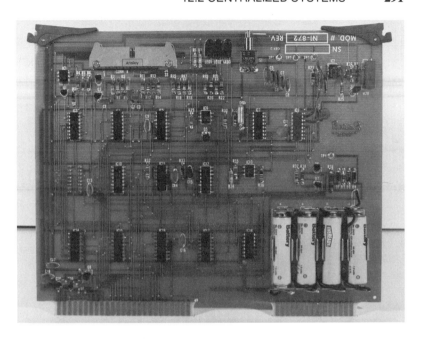

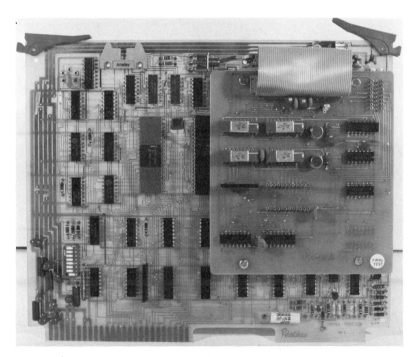

FIGURE 12-6   Much of the complex circuitry of automated systems is contained on circuit cards such as these. Equipment downtime is minimized, since entire cards can be replaced to restore malfunctioning equipment to duty. Replaced cards can either be repaired or discarded depending on the economics involved. (Courtesy of Robertshaw Controls Company.)

The programming of the simplest system is accomplished at the equipment manufacturer's plant. The system as sold has only the flexibility designed by the manufacturer. The more versatile, hence expensive, systems have the capability to be programmed either by "store-bought" prepackaged programs or by user generated programs. Such programs, referred to as "software" can be designed to meet requirements peculiar to a specific installation whereas manufacturer-generated programs are usually more general.

A typical program could provide a sequence of steps to be carried out in the event of a fire. It could trigger a tape machine to announce evacuation instructions to the fire floor and the floors immediately above and below. It could simultaneously shut off the fan on the fire floor and adjust the dampers on the floors above and below to maintain a positive pressure to minimize smoke travel. It could also dial the fire department on the telephone and announce by a prerecorded message that there was a fire and also give the location. It could then trigger the sprinkler system in the fire area, signal some elevators to go to evacuation floors and some to go to the ground floor to bring fire fighters to the scene of the fire, and also ensure that elevators would not inadvertently go to the fire floor with passengers. Finally, it could show on a central panel exactly where the flames were located as well as smoke concentrations.

The key to the ability of the supervisory system to perform any of these operations is in the programming. The central equipment is only able to do what it has been instructed to do. Usually, the programmers are not building systems people but rather technicians adept at arranging a series of instructions in a manner that will cause the central processor to do what the building operations people wish, within the capability of the "hardware" (the equipment) in the HVAC system. In smaller control systems the desired functions are built in; larger systems may also have built-in functions along with a library of other programs used occasionally. A typical program, perhaps one for scheduling preventive maintenance, may be stored on a magnetic tape similar to the cassette tapes used in tape decks. The instructions on the tape may request the CPU to scan all motors in the building, identify those that have been operating for 2000 hours since their last maintenance, and present this information in the form of a work order for lubrication service. A single cassette may contain dozens of such programs.

The HVAC operations people need not know the inner workings of the CPU. However, they must know how to make the equipment do what is desired, be able to interpret the information presented, and be able to recognize a malfunctioning control system. Detailed operating instructions as well as training sessions on the job and even at the manufacturer's plant are quite adequate to provide an interested operator with enough information to operate the system satisfactorily. As with any phase of HVAC controls, on-the-job experience will provide the bulk of the operator's expertise.

Another program of particular interest calls for the CPU to scan all the input points, several times per minute, and identify any points that are in "alarm"; that is, sufficiently removed from the desired value to be a cause for concern. In essence the CPU receives the actual value of the variable from the input device, compares it to the maximum and minimum desired values, and if not within the desired range will take some action. The action may be an output signal to a damper or valve actuator or it may be an alarm signal to the CPU operator on duty bringing to the operator's attention that something is amiss. The operator then gets on the phone and sends a maintenance person to the point of the problem. In some systems an intercom is provided so that the CPU operator can relay information to the maintenance person in a remote equipment room. This same intercom can serve as an audio monitor enabling the CPU operator to listen to motors in operation at remote locations. Squeaking belts, rumbling noises, vibrations, and other telltale indications could cause a maintenance person to be dispatched. It should be increasingly obvious that a central automated supervisory system can be a real labor saver if operated by a skilled person. It should also be obvious that an HVAC controls person rather than a "computer operator" would prove to be most valuable in operating such a system.

12.2.4 Output Devices

The reason for having automated building operation supervisory systems is to generate useful output. This output is in the form of digital or analog signals, which in themselves mean little to us humans. These signals must be converted into a form that we find useful. The programming or software accomplishes this conversion.

In order to operate the HVAC system, signals flow from the central equipment through the transmission links to the remote data collection panels, where the signals are converted from electronic impulses into electrical or pneumatic signals. Incidentally, describing the three main types of control systems earlier, it was noted that electronic systems were being used increasingly because of their compatibility with automated supervisory systems.

The automated control systems in use today do not provide functions that were not already achievable in the field with simpler equipment. Load shedding, economizer cycle, fire alarm, elevator control and so forth, are all available in simple, noncomputerized packages. The advantage of the complex system described here is

FIGURE 12–7   This central station has a CRT. Note the display of part of the HVAC system, a printer at the right that provides a record of system performance, and the central electronics cabinet at the left. (Courtesy of Robertshaw Controls Company.)

that it is centralized, subject to continuous monitoring, and provides control that can be overlapped so that if there is a failure in one area, another subsystem may take over until repairs are made. The system can also be controlled more effectively.

Beyond the output in the form of signals designed to produce corrective action, we have information signals. Data can be accumulated for study. Several times a minute all the control points in the system can be scanned. We probably do not want to know all the control point values several times a minute, so we program the computer to print out these values perhaps only once each hour. Alarm points, that is, those points with values falling outside the desired values, can be printed whenever they are noted, and in some systems they are printed in red. Newer systems use a cathode ray tube display (CRT) in addition to printing. The CRT is like a television tube, which displays numerical data much in the way a typewritten printout sheet would. Some newer systems even have color CRT capability. Most of the data scanned by the computer go into a memory unit and are never seen by the human eye. We can communicate with the computer by means of the console, a typewriterlike keyboard with alphanumeric characters, and call upon it to present the data in a number of different ways. We may want comparisons over a period of time either in the form

FIGURE 12–8   This operator's console has an intercom system enabling the operator to communicate with personnel at remote equipment rooms. (Courtesy Robertshaw Controls Company.)

of a graph or perhaps a table. We may want to study only a small section of the HVAC system and disregard the rest. Perhaps we are interested in energy consumption projections based on historical operating data. If the software has been developed and is available for the system we are using, then such information can be obtained by merely pushing a few buttons.

Troubleshooting can be handily done with such systems. First, the scanning process reveals an alarm point and displays this point on the CRT. The point code number, in essence the address of the particular point, is displayed along with its desired and actual values. By pushing another button, a diagram of the HVAC system containing this point flashes on the screen. In older systems, slides were flashed on a movie screen. Generally, about 80 slides of various subsystems could be stored. The computer memory and CRT arrangement presently coming into use can store and display upon demand many more such subsystem diagrams. On the display the location of the alarm point is shown along with actual and desired values. Other control points in the same subsystem can also be displayed so that a complete picture of temperatures, pressures, flows, and so forth, can be created. This is done before a maintenance person is dispatched to the troublespot. This is a vast improvement over having a service person attempt to obtain the same data manually. A skilled technician at the central console can troubleshoot the system and dispatch a less-skilled mechanic to make on-the-spot repairs.

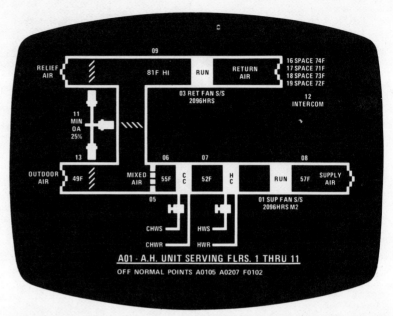

FIGURE 12–9   This CRT display shows an air handling system designated A01 appearing on the screen. The operator can look over this system and see the actual temperatures at various points. Abnormal temperatures can be displayed in color in some automated systems, or will be flashing or called to the operator's attention in some other manner. (Courtesy of Robertshaw Controls Company.)

In addition to printers and CRT display, output data can also be recorded for storage outside the limited memory bank of the CPU. Such data can be recorded on magnetic tape, discs, drums, chips, and other storage devices that are being developed daily by the rapidly changing information management industry.

12.3 A TYPICAL SYSTEM

Just as the HVAC field has a unique vocabulary, so does the computer field. When the two fields are joined together, as occurs in building operations supervisory systems, the problem is compounded, particularly when manufacturers use terms peculiar to their systems only. In this section we consider features as they are described by one manufacturer and attempt to relate them to the general descriptions given previously.

The following are descriptions of features found in a brochure presenting a general purpose central monitoring and control system for building automation, energy conservaticn, and fire and security.

1. System Operational Modes: Local-Remote, Night-Day, Auto-Manual, Normal-Maintenance. (These modes of operation are selected at the option of the console operator.)

2. Software. Fixed user software (prepared specifically for the customer to accomplish particular tasks that are built into the CPU), variable field user software (programming that can be modified in the field by the console operator depending on the operating situation encountered), and a library of application programs.

3. Analog System. Full digital transmission with A-D or D-A conversion. (This indicates that either digital or analog sensors can be used and that the signals will be changed to digital form for transmission and transformed back to analog form if necessary.)

4. Powerful set of system commands, 50 custom user-commands also allowed. (The console can be made to contain a number of programmed commands that are put into use by depressing only one or two buttons in addition to the normal complement of keys on the console.)

5. Point Selection. Alphanumeric keyboard of the operator's video terminal. (Control points can be selected for view on the CRT by pressing a series of letter/number keys.)

6. Control Point Adjustment. CPA and alarm limits entered directly in engineering units. Alarm limits slide with a change of CPA. [Control point values can be changed from the CPU console. This is done in °F, gpm, cfm, psig, and so forth (engineering units) rather than some mysterious code. The maximum and minimum allowable values will automatically be changed when the control point is changed.]

7. Logs. All-point, trend, totalizer, alarm summary, status, and start–stop schedules. (These logs are printout records that can be kept and presented both on a printed form and on the CRT at the push of a button.)

8. Alarms. Audible alarm, visual display on the operator's video terminal, and printout. Acknowledgment is required. The time and date record of the alarm, acknowledgment, and return to normal are included. (The console operator must tell the CPU that he is aware that an alarm condition has occurred.)

9. Hardware Malfunction Monitoring. Autoreporting of power failure, system failure, hardware malfunctioning, and line tamper.

10. Software Malfunction Monitoring. "Dead-man stick" program for the autoreporting of software failure or crash. (In parts (9) and (10) equipment or program failures will automatically notify the operator of problems and put the system in a standby mode.)

11. Maintenance Mode. Special diagnostic software for system debugging, startup, checkout, and ongoing preventive

maintenance. (Problems in the automation system itself are continually sought and reported.)
12. Training Mode. Special software for rapid operator training, on-site, under actual conditions of system operation.

These features are part of the system made up of so-called building block components described as follows:

Data Panel

Located at strategic points within the premises, connected to sensors (digital and analog) and control devices (on-off, set point adjust). Exchanges monitoring and control information between the sensors/control devices and the Central Console using carrier communication on existing power lines or dedicated two-wire lines. (This is the remote data collection panel.)

Communication Processor (Central Console)

Receives and processes all emergency and status information, continuously supervises and controls the entire system, and implements operator commands. It includes a central microcomputer, system interface, video terminals and alphanumeric printers (on-site and off-site), computer peripherals, and graphics projectors. The processor operates stand-alone as a Central Console or as an adjunct to a master computer.

Operator's Video Terminal

Man/machine interface designed for ease of system operation and fast human response. The terminal includes an English language CRT display and alphanumeric keyboard.

Master Computer

Expands the system capability and allows system configuration for large regional networks. It communicates with slave Communication Processors over telephone lines and can also be used to provide redundant Central Console backup, for fail-safe system operation.

Peripherals

Includes wide carriage printers, video terminals, slide (graphics) projectors, paper tape reader, magnetic tape cassette, and disk cartridge.

Special Optimization Programs

In addition to the wide range of user software provided with the system, the following special programs are also available: fire alarm/security program, audio control/annunciation, patrol tour, remote meter reading, time-of-day metering, submetering, electric energy profile, demand management, time/event load control, programmed lighting control, programmed start/stop, supply air reset, enthalpy management, chiller profile management, boiler profile management, maintenance operation, totalizing of parameters, and energy management information system.

To those uninitiated in the computer field much of the foregoing may seem foreign and have little meaning. It should serve as an introduction to an area of HVAC still in its infancy. We can see how powerful a tool automated building operations supervisory systems can be. Like any tool, however, it requires an operator skilled in its use. A high-powered tool in the hands of an incompetent operator is a waste, pure and simple. Many automated systems in use today are underutilized and are not realizing their full payback potential. The skilled controls technician has a fine future in this growing field.

DISCUSSION TOPICS

1. List several major jobs done by a supervisory system.
2. What are the factors usually studied in deciding that an automated system will be installed?
3. Give some examples of input devices. Output devices.
4. What is the purpose of the remote data collecting panel (RDCP)?
5. What is a CRT? Where is it used?
6. What is a modem? Why is it needed?
7. What is software? Give several examples.
8. How would you design an automated system for your home?

Chapter Thirteen

Electrical Troubleshooting

When a brand-new HVAC system is to be started up for the first time, the controls person skilled in troubleshooting should be in attendance. Why? Because in more cases than not the system will not work. The appropriate buttons will be pushed and all present will wait for things to begin to happen, but unfortunately something will not work. What do we do now?

13.1 KEEP CALM

The first step in any troubleshooting exercise is to become familiar with the system and sequence of operation. System problems (except by luck) cannot be found without a good idea of *what should happen* when a particular button is pressed. However, with the theory of operation in mind, we can progress to the next step. This is a physical inspection of the system to identify the actual hardware and to relate it to the wiring diagrams and sequence of operation under study.

As our field experience increases, the inspection and orientation time decreases. On any new system, though, it is always a good idea to get the total picture before proceeding to the fine details.

Next we make sure that the plug is in the wall. In large central systems this is the equivalent of checking to see that proper power is present at the power panels. Whether a system is new or old, we need to check for power and *measure* its voltage with a voltmeter. Improper voltage has kept more than one system from starting.

Then we check the reset buttons on all the manual reset safety devices. They may be in the tripped position due to rough handling in shipment or installation. In older systems they may, in fact, be tripped because of an overload situation. By following these simple steps before attempting to use diagnostic instruments a substantial amount of effort might be saved.

At this point, if there is still no success, it is time to resort to test instruments.

13.2 INSTRUMENTS

There are many types and sizes of electrical test instruments on the market today. The ones that are of primary importance, and will be dealt with here, are the VOM and the ammeter. These are the instruments that will provide actual quantitative values for voltage, resistance, and current draw of electrical devices. They are all one needs to be an effective troubleshooter. All other instruments, with perhaps one or two exceptions, are variations and combinations of these basic tools—some dressed up, some stripped down, some with digital readouts, some with flashing lights. They may be fun and convenient but they are not essential.

Many troubleshooters in the field today use a test light, jumper wire, and continuity tester and rarely, if ever, pick up a meter. We shall consider these devices, but it is likely that these troubleshooters, as good as they may be, could be even better using a meter.

13.2.1 The VOM

The volt–ohm–milliammeter, or multimeter as it is often called, is a versatile test instrument capable of measuring ac and dc voltages, resistance in ohms, and current draw in milliamperes. It is available in a range of prices from under $20 to over $200 depending on the range selection, durability, ease of reading, accuracy, and so forth. As in everything there is a range in quality in these meters also. Measuring dc voltages and reading milliamperes current is useful in electronic systems. Measuring ac voltages on scales up to 600 volts, or more, for line voltage equipment and down to 50 volts or less for low voltage control circuits provides all the flexibility we need for ac circuit analysis.

The ohmmeter scales enable us to get resistance readings of less than one ohm up to several hundred thousand ohms. We can check out heaters, relay coils, motor windings, and switches handily with this capability.

In most field troubleshooting jobs it is not necessary to have high accuracy in a meter. Accuracy within about 2% of full scale is usually quite adequate and can be obtained fairly inexpensively. An accuracy of ¼% is costly and usually reserved for the test laboratory. Field instruments should be able to take rough handling, be handy to use, and give accurate information.

Since the voltmeter measures electrical potential, the system being analyzed must be energized. That does not mean necessarily that it has to be running, but at least power must be applied to the system. The ohmmeter, on the other hand, generates its own potential with a built-in battery. It is designed to test deenergized (no power applied) systems. In fact, if one attempts to test a "hot" system, one can burn out the meter movement on some meters. Better meters have a replaceable fuse that will blow, saving one the embarrassment of explaining how one destroyed the meter.

Because there are some variations in connecting different VOMs, it is necessary to read the manufacturer's instructions before attempting to use them. In Figure 13–1 the selector knob of one type

Highlights—The Lockout Relay

Manual reset safety controls are used in order to draw the attention of the equipment owner to the fact that an unsafe condition exists in the equipment. If the equipment is located on the roof or in a remote location, the resetting of manual controls can be a chore and even require a service call. This can be costly, particularly if the condition that caused the safety to trip was a random one and not indicative of an equipment shortcoming. This problem can be minimized by using a "lockout relay" or, as it is called by some manufacturers, a control relay.

Figure 1 shows a portion of a more complex wiring diagram for a rooftop air conditioning unit. The circuit shown is for the coil of a compressor contactor (CC). Under normal operating conditions, when the thermostat calls for cooling, the compressor relay (CR) coil is energized (not shown) causing the CR contacts to close. The compressor contactor coil will be energized through the circuit containing the overload (OL), low pressure cutout (LPCO), high pressure cutout (HPCO), and lockout relay contacts (LR).

Observe that the current flow through the lockout relay coil (LR) is essentially zero, as the electricity takes the path of least resistance through the closed switches to energize coil CC.

Now assume that an unsafe high pressure has caused the HPCO to open. Current will flow through coil LR and coil CC. Since there are now two coils in series, the voltage drop across them will be related to the resistance of each. The voltage drop across coil CC will no longer be enough to enable it to hold in the compressor contacts. The compressor will shut down.

The voltage drop through coil LR, although less than 24 volts, will be enough to cause it to open the normally closed contacts LR. At this point the compressor is "locked out." Even if the HPCO switch closes as a result of decreasing pressure, the LR contacts will remain open and keep coil CC from being energized.

To get the compressor going again, the equipment owner need only walk over to the thermostat and turn it to a higher setting to deenergize the compressor relay CR. The CR contacts will open and deenergize the circuit permitting contacts LR to close. When the thermostat is turned down to its original cool setting, the CR contacts will close and the compressor coil will be energized once again.

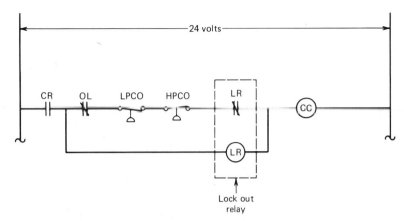

FIGURE 1

If the condition causing the HPCO to trip has not been corrected, the cut-off and lockout cycle will repeat. Once the problem has been corrected, the equipment will run normally. If the owner of the equipment has to go through the resetting procedure regularly, it should be obvious that a problem exists that needs attention. The lockout relay is widely used as an alternative to manual reset safety controls.

FIGURE 13–1   The versatility of the VOM is reflected by the variety of settings obtainable with the selector knob. The meter should always be "zeroed" prior to using to insure accuracy. It is important to remember that the ohmmeter is used only when the electrical circuit is deenergized. (Courtesy of Simpson Electric Company, Elgin, Illinois.)

is shown. When measuring an unknown voltage, it is a good idea to place the selector at the highest setting and slowly turn the knob toward the lower settings until the needle lies in the middle half of the scale. It is in this range that the meter is most accurate. The increments between the numbers on the various scales may be different so that it is necessary to be careful when reading the voltage indication to be aware of the size of the increments being read. The knob position and the scale should be checked continually to avoid error. If there is a major shortcoming in using the VOM, it is the possibility of becoming confused and reading the wrong scale.

When using the voltmeter, it is a good idea to take the voltmeter leads and measure a hot circuit such as a 120-volt wall outlet to make sure the meter is working. This should be done just before measuring the unknown voltage so that there is no danger of being electrocuted by touching live wires a defective meter failed to detect.

Figure 13–2 shows a voltage reading being taken. The meter leads are held by the insulated grips, and the bare metal tips are held against the terminals being measured. Enough pressure should be applied to get a good metal-to-metal contact. The metal tips can be rubbed against the terminals to scratch through any dirt or oxide buildup that would prevent a good reading.

FIGURE 13-2   The VOM is a convenient instrument for checking the availability of power. The instrument should always be checked on a known power source before being used to make sure that it is working properly. One's life can depend on the reliability of this meter. If a person touches "hot" wires that tested "dead" with a defective meter, the results might be terminal.

CAUTION: Make sure your fingers that are not holding the insulated grips do not stray against hot leads. Do not bend your neck in an effort to see better and bring your head into contact with hot parts. People have been injured and, from time to time, killed by such action. Be *respectful* of the power of electricity and be *cautious.* Accidents are *wasteful.* As you take the reading, also be sure that the two metal tips do not accidentally touch each other or other metal parts, as they will create a short circuit with accompanying sparks that could cause damage. This is particularly true when troubleshooting some control panels with very close quarters (which for some reason most of them have).

In using the ohmmeter to measure resistance, the equipment must be deenergized. If the range of the resistance to be measured is not known, we start at the lowest resistance scale, $R \times 1$, and proceed to progressively higher scales until the needle registers somewhere toward midscale. With very low or very high resistances this may not be possible. The lowest resistance we can read is zero, indicating no resistance. Theoretically this may not be possible but in field measurements it is commonly seen when resistance through switches and wires is measured. Resistance through relay coils and heaters may be in the hundreds, $R \times 100$, or thousands, $R \times 1K$, ohms. Resistance across capacitors is measured in the hundreds of thousands of ohms, $R \times 100 K$, and

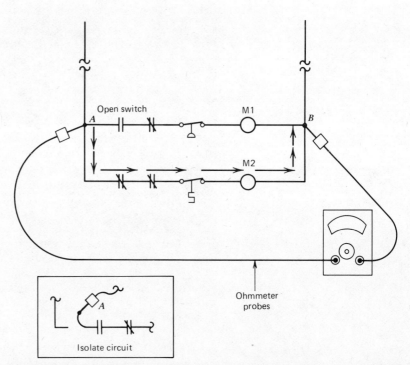

FIGURE 13–3   In measuring continuity between A and B, there are two paths the electricity can take. The open switch will not be detected if the ohmmeter is used as shown here, since the electricity will bypass the open switch and yet show a continuity reading. Isolating the circuit by removing the M1 circuit from the rest of the system, as shown in the inset, will insure a true reading.

an open circuit caused by an open switch or broken wire will show an infinite reading (∞).

It is good practice to remove a wire lead from one end of a circuit or from one terminal of the circuit or device being tested for resistance. (See Figure 13–3.) Otherwise, under certain circumstances, an incorrect reading can be obtained. We may think we have a continuous circuit when in fact we are reading continuity through an adjacent circuit. By isolating the circuit under test we cannot make such an error.

13.2.2 The Ammeter

An ammeter must be connected in series with the load being tested. When using a VOM, the milliammeter is connected in series. A milliampere is one thousandth of an ampere; since most VOMs have a maximum capability of about 1000 milliamperes (ma) or 1 ampere, the meter is quite limited in use. It is adequate for many electronic applications but for few electrical power or control circuits. The instrument used in the field in place of the VOM is called a wrap-around ammeter. This instrument has a set of metal jaws that can be used to encircle a conductor. The current flowing through the conductor generates a magnetic field of an intensity proportional to the current. The wrap-around ammeter measures the intensity of this field and gives a readout in amperes.

Some of the field instruments combine current scales with voltmeter and ohmmeter scales to make a very versatile instrument. Because the resistance scales are usually rather limited and handling the wrap-around meter may be awkward at times, it is useful to have both the wrap-around meter and the VOM available.

In Figure 13–6 the meter is being applied to a conductor to measure amperage. There are usually a number of scales covering a wide range of current ratings. We select the maxium scale to prevent damage to the meter when measuring an unknown current. We gradually shift to lower scales until the meter indication is roughly at midscale for maximum accuracy.

In the event that the anticipated current draw is higher than the maximum scale available, there is an accessory device used to step down the current reading by a factor, say of 10, so that a measured current draw of 200 amperes would actually be read on the meter as 20 amperes. Similarly, stepup devices are available that multiply the current draw by a factor, say 5 for example, so that a current of 0.3 amperes would be read as 1.5 amperes on the meter. Dividing the reading by the factor gives the proper amperage.

In control circuitry it is not uncommon to have currents that are quite small. In Figure 13–7 we see a trick used in the field that multiplies the current read by the meter without the use of an

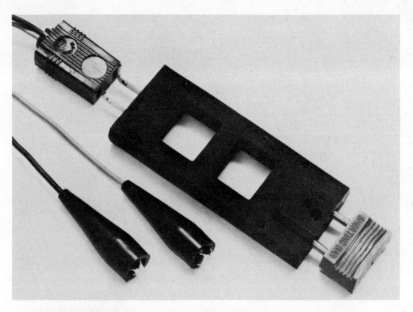

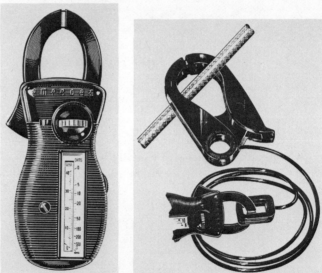

FIGURE 13–4   The wraparound ammeter, shown at lower left, is a vital test instrument for the controls person. When used with the "energizer", shown at the top, to step up very small current readings; and with the "decatran" at the lower right, to step down very high current readings, it becomes a very powerful diagnostic tool. (Courtesy of Amprobe Instrument, Division of Core Industries Inc.)

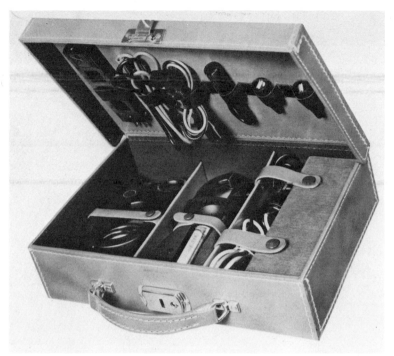

FIGURE 13–5   A wraparound ammeter kit containing the basic meter as well as accessories that enable very low and very high current readings to be obtained using the existing range of the meter. Various test leads to facilitate the use of the instrument are also provided. (Courtesy of Amprobe Instrument, Division of Core Industries Inc.)

accessory device. The conductor is wrapped around the jaw of the meter perhaps 10 times. This multiplies the current by 10 so that a 1-ampere reading is actually only a 0.1-ampere current. This is quite handy in setting up a heat anticipator in a thermostat or measuring low values of current in commercial control circuits.

In using the meter, we make sure that only one conductor at a time is being measured. A lamp cord such as that shown in Figure 13–8 has two conductors insulated and bonded together. The insulation is not a problem; however, the conductors must be split apart and only one conductor at a time measured. The effect of measuring both conductors is to have the current in one cancel the current in the other giving a zero reading. In Figure 13–4 the "energizer" is used to split the current through such a cord. It is quite handy for small equipment with line plugs. In larger equipment, though, a spot must be found where each conductor can be measured separately, usually close to the load, or in a control panel.

FIGURE 13–6   The wraparound ammeter is handy for measuring the current draw in control panels. One should be careful, when using the meter, to avoid inadvertently touching adjacent wires with a bare hand or ground-out wires with the meter. In electrical control panels of the type shown, this is quite easy to do.

13.2.3 Jumpers

A jumper is nothing more than an insulated piece of wire, or a screwdriver with *insulated handle,* or a pair of pliers with an *insulated handle,* placed across a switch to see if the circuit is open. If the load is energized with the jumper in place, the switch is open; if nothing happens when the jumper is applied, presumably the switch is closed. This is a rather imprecise technique but one

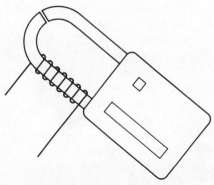

FIGURE 13–7   Wrapping the conductor around the jaw of the ammeter multiplies the current reading by a factor equal to the number of wraps. An actual current reading is then obtained by taking the displayed reading on the meter and dividing by that number of wraps.

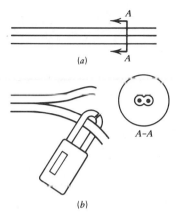

FIGURE 13–8   A lamp cord (*a*) that has two conductors insulated together (see Section A-A). In order for the wraparound ammeter to be used, the conductors must be split apart and only one conductor should be measured.

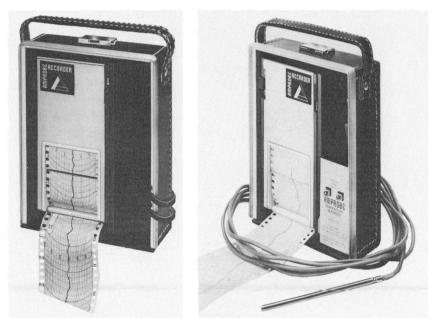

FIGURE 13–9   Two examples of recording instruments that are very valuable in tracking down control system problems caused by transient conditions or that may occur when no one is around to observe them. The records provided by these devices may save many hours of on-the-job observation in troubleshooting. (Courtesy of Amprobe Instrument, Division of Core Industries Inc.)

that is very widely used. Its main advantage over a meter is speed: We can test a number of switches very quickly. By merely touching the jumper to the switch for an instant, listening for a reaction (i.e., does the motor run or does the relay pull in?) and then quickly pulling the jumper away, we can determine if a switch is open.

A jumper cannot tell us if voltage is present. It also cannot tell the magnitude of the voltage. It can cause sparks to fly; it can become welded to the switch terminals (screw drivers and pliers have been ruined by the melting action); and, in short, it can be hazardous to our health and can damage the equipment being serviced. In the hands of an experienced service person who knows exactly what is to be done, it can be a useful tool but it is of limited value and certainly not a replacement for a meter.

A test light can be used for determining the presence of voltage. A neon lamp tester can distinguish 120 and 230 volts by its intensity although no exact indication is given. The test light can be used to troubleshoot energized circuits in the same manner as a voltmeter, but generally it gives a ''go-no go'' indication rather than quantitative data. The experienced service person can find this quite adequate in many applications, but it is inferior to a meter in many others.

A continuity tester generates a small potential with a battery and gives an indication with a light bulb. Some flashlights have built-in continuity testers. They are quite adequate for testing many circuits with switches. They do not give a resistance reading in ohms, which is important in checking loads like relay coils and motors; however, they are usually reliable in checking circuits containing switches just to see if the switches are open or closed. It is a durable tool for making quick checks but does *not* replace the ohmmeter.

13.3 TROUBLESHOOTING —CASE I

The wiring diagram in Figure 13–11 is for a residential heating/cooling system with a 230-volt condensing unit, a 115-volt evaporator/furnace blower motor, and a 24-volt control system. Prior to making any effort to troubleshoot the system, a tour of inspection is necessary. Note the 115- and 230-volt power supply, the transformer, and the location and identification of the various loads. In the actual installation the room thermostat may be in the upstairs hallway, the furnace in the basement, and the condenser in the backyard. It all looks very neat on paper, however.

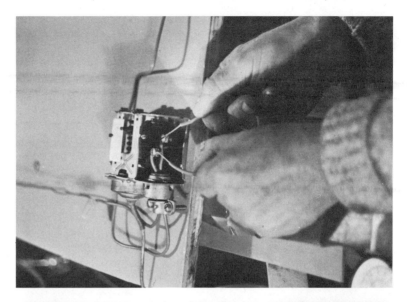

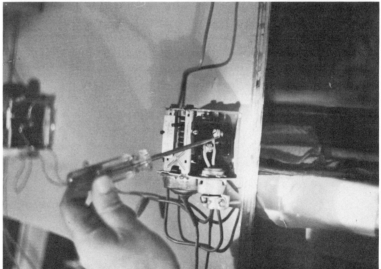

FIGURE 13-10   The jumper wire and screwdriver blade, handy tools for checking whether a switch is open or not. Many mechanics have never passed beyond this stage of troubleshooting. In the hands of someone who knows what he is doing, the jumper can be quite useful. *If used carelessly it can be hazardous to one's health.*

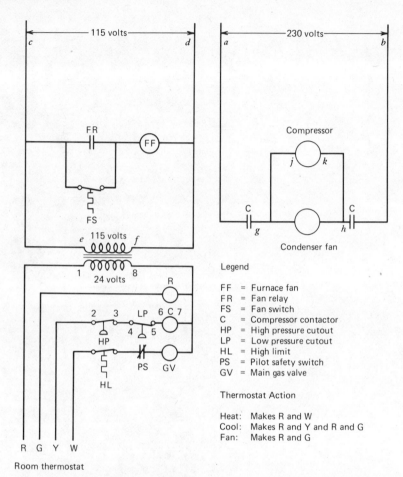

FIGURE 13–11   A typical diagram found in residential air conditioning and heating systems. Note that the thermostat switching has not been shown in the schematic but is described in the legend.

Problem 1.

The compressor and the condenser fan do not operate. The furnace fan does operate on a call for cooling.

    a. After a visual inspection, check for the availability of power at points *a* to *b*, *c* to *d*, and *e* to *f* using the voltmeter. Insure that the voltage readings are proper, *a* to *b* = 230 volts, *c* to *d* = 115 volts, and *e* to *f* = 115 volts.

    b. Insure that the control circuit has power by checking for 24 volts across points 1 to 8 with the voltmeter.

    c. Determine which circuits may be at fault. Since both the compressor and the condenser fan are inoperative, a check at points *g* to *h* with the voltmeter should give a zero reading. That being the case, the switches of the compressor contactor, labeled

C, are probably open. This will happen if the coil of the contactor is not energized.

d. Check for voltage at 6 to 7. A 24-volt reading would indicate that the coil is defective or perhaps there is a mechanical problem with the contactor that prevents it from pulling in. In any case a 24-volt reading should cause the contacts to close, and failure to do so requires that we inspect the compressor contactor further.

A zero reading across 6 to 7 indicates that the control circuit is broken and further investigation is necessary. A procedure called "hopscotching" will be used.

e. With the system still energized, we place one lead (either one) of the voltmeter at 1. The room thermostat is a temperature-actuated switch. We must know the switching action before attempting to troubleshoot. Experience will teach us many typical

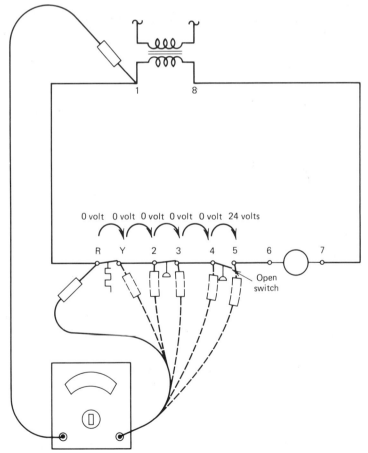

FIGURE 13–12   "Hopscotching" is a useful troubleshooting technique that can quickly identify an open switch. In this cooling circuit, isolated from Figure 13–10, one can quickly determine that the LP control is open.

switching actions. Manufacturer's specification sheets also show this action. (See Figure 13–13.) The legend of the wiring diagram shows that on a call for cooling, the switch between *R* to *Y* in the thermostat will be closed. Placing the other lead of the voltmeter at terminal *R* should give a zero reading which indicates there is no break in the line between the transformer, point 1, and the thermostat terminal *R*.

f. Continuing to hold one lead at 1, we move the other lead to *Y*. A zero reading indicates that the thermostat switch is closed. A 24-volt reading indicates that the switch is open; that is, either the thermostat is not calling for cooling *or* the thermostat is defective.

g. Now we move the voltmeter lead to point 2. A zero reading shows that the wire is intact from *Y* to 2; a 24-volt reading indicates that it is broken.

Q539 THERMOSTAT SUBBASES

PROVIDE MOUNTING AND MANUAL SWITCHING FOR T87F THERMOSTATS.

Include cooling anticipator, letter-coded and color-coded screw terminals for electrical connections, and fingertip control of fan and system switch levers. The wide range of switching functions fits most cooling or heating-cooling applications. Electrical Ratings: 2 amp at 24V ac. Optional indicator lamp: 28V ac. All Q539 subbases mount directly on wall. To mount on an outlet box, use 6 in. [152.4 mm]

cover ring (included with some T87F models) or order 129044A Adapter Plate Assembly. Approximate Dimensions (with T87F): 3-11/16 in. [93.7 mm] dia., 1-3/4 in. [44.5 mm] deep.

TERMINAL DESIGNATIONS:[a]

TERMINAL	CONNECTION
R	Transformer.
R_H	Heating transformer with isolated heat-cool circuits.
W	Heating relay or valve coil.
Y	Cooling contactor coil.
B, 4	Heating damper (if used). Circuit only completed between R and B with system switch in heat position; "4" terminal is the same as Honeywell "B" terminal.
O	Cooling damper (if used). Circuit only completed between R and O with system switch in cool position.
G	Fan relay coil.
X	Clogged filter switch. Only available on subbase with malfunction light.
P	Heat pump contactor coil.
Z	Q539D,H—low voltage fan switch for control of fan relay in AUTO position for both heating and cooling control.

[a]R_1, W_1 and Y_1, are not marked on thermostat subbase. They are mounting posts and electrical connections for the thermostat.

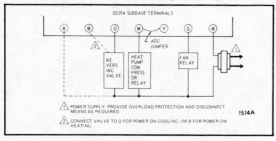

Typical Q539A-T87F heat pump hookup.

FIGURE 13–13   Part of a controls manufacturer's specification sheet depicting the internal switching action of a thermostat as well as giving terminal designations. Although wiring arrangements frequently encountered can be quickly committed to memory, such sheets are very helpful in understanding infrequently encountered controls. (Courtesy of Honeywell Inc.)

h. Then we move the lead to point 3. A zero reading means that the HP switch is closed, a 24-volt reading means that it is open. We continue to move the second lead from point to point (hence the name hopscotching) until we get a 24-volt reading. Let us assume that happens at point 5. Since everything was intact up to point 4 and is now open at point 5, the problem must be that the LP (low pressure) switch is open. It may be defective or perhaps (and most likely) the refrigeration system needs refrigerant; in any case attention is now focused on the LP switch for further analysis.

This looks quite neat and, in fact, many troubleshooting procedures can be done as handily as just shown. In the example described, it cannot be as neatly done as indicated because the room thermostat, transformer, and other controls are widely separated. We just cannot extend the meter leads or stretch our arms that far. We have to modify the procedure a bit.

After determining that we have control system voltage at points 1 to 8, we walk up to the hallway with the thermostat and remove the cover exposing terminals R-G-Y-W. Then we place the voltmeter across R to Y. A voltage reading indicates that they are open; a zero reading indicates that they are closed. Now we move outside to the control panel containing the HP switch and put the leads across 2 to 3. Again *voltage means open; no voltage means closed.* The value of hopscotching is that it forces one to be systematic. If it is not feasible, then checking out each component individually, in some organized way, will give the same result.

REMEMBER: In an energized circuit a voltage reading across a *switch* usually means it is open, a zero reading usually means it is closed. A voltage reading across a *load* means it should operate; a zero reading means it is deenergized and should not operate.

Problem 2.

The furnace fan, compressor, and condenser fan fail to operate on a call for cooling.

a. After visual inspection, we check the voltage at points *a* to *b, c* to *d,* and *e* to *f.* We should get 230, 115, and 115 volts, respectively.

b. Ensure that the control circuit has power by measuring across points 1 to 8 wth the voltmeter. We get a reading of zero.

The diagram shows a transformer with 115 volts going into the primary and 24 volts coming out of the secondary. The voltmeter reading of zero out of the transformer with proper voltage going in means that the transformer is defective. With no control circuit power, none of the relays and contactors can work to close the contacts and energize the motors.

Problem 3.

The condenser fan and furnace fan work, but the compressor motor does not on a call for cooling.

In checking the wiring diagram, it should be apparent that both the condenser fan motor and the compressor are intended to operate together. If one is working and the other is not, we can surmise that the control circuitry is not at fault because switch C must be closed. We now check the load, the compressor motor first.

We place the voltmeter leads at points j to k and read 230 volts.

Since the load is energized, it should work. Since it does not, it must be replaced or repaired.

As you can see, some problems are very easily solved while others require a lot of effort. Some problems are never really solved because they are intermittent in nature and invariably when the service person is examining the "patient" all is well. As soon as the service person leaves the job site, the problem recurs. In such situations you guess or wait until the problem becomes sufficiently obvious so that it can be readily solved. The key to troubleshooting, assuming of course that we already know the equipment and the principles of operation, is *patience* and *persistence*.

13.4 TROUBLESHOOTING —CASE II

The wiring diagram in Figure 13–14 is for a rooftop gas-fired heating unit with electric air conditioning serving a restaurant. A cursory inspection of the diagram shows that three-phase power supplies the system. There is a transformer, which is unmarked but has a line voltage primary side, and we can surmise 24 volts on the secondary side. Although this is actually true, we cannot really tell from the diagram. Eventually, when we get to the equipment, we shall have to look at the transformer and read the fact that the secondary voltage is 24 volts from the nameplate data.

Looking at the control circuitry, we see such things as a pilot igniter, a fan time delay relay, a compressor contactor coil, and a fan relay coil. Scanning to the left, we see four switches, seemingly manual, connected to terminals labeled 4, A, B, W, and so forth.

The table off to the left indicates which of the four switches are closed during different modes of operation. Anyone with no prior field experience, no written sequence of operation, and no knowl-

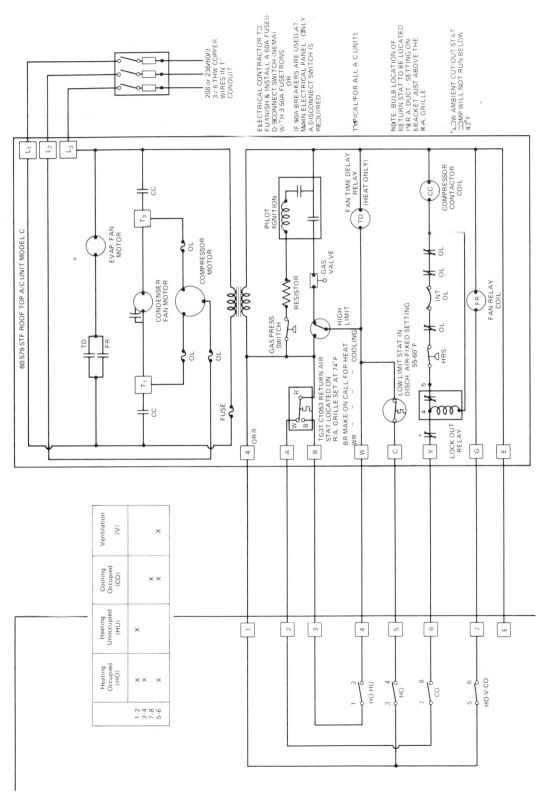

FIGURE 13–14 This rooftop air conditioner wiring schematic is difficult to interpret unless the operator has had a good amount of field experience or has available the sequence of operation to read as the diagram is being studied.

edge of the hardware used in the field, will be hardpressed to understand this diagram. Still, in order to learn, it is necessary to forge ahead.

Problem 1.

The evaporator fan motor does not run during a call for cooling.

a. Visually inspect the unit identifying parts and components.

b. Measure the incoming power with the voltmeter at points L1 to L2, L2 to L3, and L1 to L3. All three readings should be 208 volts.

c. The evaporator fan motor is a single-phase motor wired between L1 and L3. It is a simple circuit with two switches wired in parallel such that if *either* TD (for time delay relay) *or* FR (for fan relay) is closed, the motor will run. We measure the voltage at the motor. We assume a reading of zero, which means that the motor is not energized and should not be running.

d. Since a relay must be energized (either TD or FR), we check the transformer secondary to make sure that the control system has 24-volt power. We assume a reading of 24 volts which is good.

e. Check the coil labeled TD. Notice a note indicating "Heat only." Since we are in the cooling cycle, we can ignore TD, which is not supposed to be working.

f. Check the coil labeled FR. We put the voltmeter leads across the terminals bringing power to FR. It reads 24 volts. Here is the problem. The relay is energized, but the contacts do not close; therefore, the fan motor cannot run. We replace the relay.

Problem 2.

The evaporator fan works, but the compressor and condensor fan do not on a call for cooling.

Looking at the upper part of the diagram, we can see that the compressor has a three-phase motor and the condenser fan has a single-phase motor wired across L1 to L3. Notice that only two switches, labeled CC, are used to control both motors. In three-phase circuits we need only break two of the three lines to shut down the motor. This is not a good idea from the viewpoint of the service person but not at all uncommon in practice.

In Figure 13–15 the control circuit that is to be energized during the cooling cycle has been traced out. This is actually a major step in the troubleshooting process and must be accomplished before any *systematic* analysis can be made.

The word *systematic* is underlined because occasionally a random check of a part will show it to be defective and result in a

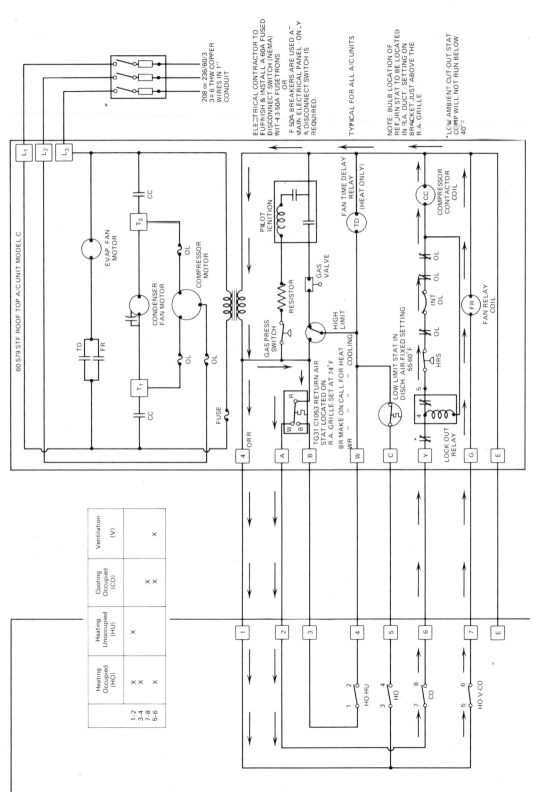

FIGURE 13-15 The two 24-volt control circuits that are energized during the cooling cycle are indicated with arrows. Notice that although nine switches must be closed for the compressor to operate, only one must be closed for the evaporator fan to be energized.

lucky "fix." To work systematically requires full knowledge of the circuitry.

Since the evaporator fan is working, we can ignore its circuit. Our interest is in the compressor contactor coil circuit. We make a quick check of the voltage at the coil—a zero reading! The coil is not energized; therefore, the compressor and condenser fan motors will not run.

Starting at the left side of the transformer and tracing out the circuit shows that there are *nine* switches, including the return air and low ambient thermostats, that must be closed in order for the contactor coil to be energized.

In the diagram the switches are neatly shown next to each other. This is the beauty of the ladder schematic: It lets one rapidly determine the switches affecting the operation of the load. Now we must locate each component within the heating/cooling unit and must check each one with the appropriate meter. The return air thermostat is sitting in the return air duct; the low ambient thermostat is on the roof in the condenser section; the three overloads (OL) are mounted on the body of the compressor; the internal overload (Int OL) is within the compressor and has two pins sticking out under a control cover; the lock-out relay is in a panel next to the evaporator fan; and the switch labeled 7 to 8 is located in a control panel in the restaurant being served by the equipment.

The voltmeter can be used with the unit energized. Measuring the voltage across each switch until we identify one that gives a 24-volt indication will enable us to pick out the open switch. Occasionally, two switches in a circuit are open. Hopscotching and checking the voltage across each switch will not locate the problem. Checking the voltage as a first approach is still a good idea, since we do not really know that more than one switch in the circuit will be open and usually this is not the case. If we suspect that it is the case, however, we *deenergize* the system and use the ohmmeter to check each component. We remove one lead from the switch as we check it with the ohmmeter to keep from reading continuity through an adjacent circuit and reaching a wrong conclusion.

Problem 3. Nothing Runs

Whenever nothing is running, there is a good chance that the power supply is at fault. Checking the voltage across L1 to L2, L2 to L3, and L1 to L3 is always a good idea no matter what the problem may be. In this case let us say that L1 to L2 = 120 volts, L2 to L3 = 120 volts, and L1 to L3 = 0 volt. The problem appears to be with the incoming power, most likely the fuses.

We check the incoming power on the line side of the fuses: L1 to L2 = 208 volts, L2 to L3 = 208 volts, and L1 to L3 = 208 volts. We know now that the problem must be with the fuses. Should we replace all of them? We could. We could also take all three of them out and check them for continuity with an ohmmeter. Those giving a zero resistance reading are good; those with an infinite resistance are blown. Finally, we could check the fuses with a voltmeter while they are still connected and energized. We put one lead of the voltmeter on the *load* side of the fuse being tested and the other lead on the *line* side of the adjacent fuse. A voltage reading of 208 volts shows the fuse to be good whereas any other voltage reading indicates that it is not good. This can be a time saver because some fuses can be difficult to remove from the fuse holder to test and replace.

Many things can go wrong with a control system depending on its complexity. The voltmeter and ohmmeter will enable us to track down most of them. The ammeter enables us to get a reading on the current draw of a load and to determine if the load is working properly. A pump motor drawing more current than the nameplate rating will overheat and burn out. A fan motor drawing much lower than the nameplate rating is not working hard enough. It may be improperly set or restrictions in the duct system may be preventing it from moving the right amount of air. Instruments can be useful in solving electrical problems and can also provide clues that will help in solving mechanical problems as well.

The skills a successful troubleshooter possesses include the ability to use instruments, to read wiring diagrams, and to draw a wiring diagram in the absence of one from the manufacturer. Also essential are an understanding of control components and how they work, knowledge of the sequence of operation of a variety of common HVAC systems, and the ability to read instructions and follow a sequence of logical steps. These skills are all possible with *patience* and *persistence*.

DISCUSSION TOPICS

1. What is the first step in any troubleshooting exercise?
2. Why is a meter superior to a jumper in troubleshooting? Under what conditions might a jumper be superior to a meter?
3. How would you check a voltmeter to make sure that it is operating properly?
4. What resistance would you expect to measure through a closed switch? Through an open switch?
5. How does a wrap-around ammeter work?
6. How can you multiply current in a wire so that the wrap-around ammeter can give an accurate indication in low current circuits?

7. If the system in Figure 13–11 were in the heating mode and the thermostat were calling for heat, what voltage would you expect to measure across the main gas valve? Across the pilot safety-switch?

8. In Figure 13–15 how would you determine if the evaporator fan motor was overloaded?

9. In the same figure, what is the function of the low limit thermostat?

Chapter Fourteen Glossary

In order to function in any technical area, it is important to be able to communicate effectively. The terminology used in controls work is sufficiently unique to warrant inclusion in this text as a separate chapter.[1] Although some expressions may vary with different manufacturers, this glossary is quite complete and accurate and reflects the terminology in use throughout the industry today.

ACTUATOR—a controlled motor, relay, or solenoid in which the electric energy is converted into a rotary, linear, or switching action. An actuator can effect a change in the controlled variable by operating the final control elements a number of times. Valves and dampers are examples of mechanisms that can be controlled by actuators.

ADIABATIC PROCESS—a process in which there is neither loss nor gain in total heat.

ADJUSTABLE DIFFERENTIAL—a means of changing the difference between the control cut-in and cut-out points.

AIR CLEANER—a device designed for the purpose of removing air-borne impurities such as dust, gas, vapor, fume, and smoke. (Air cleaners include air washers, air filters, electrostatic precipitators, and charcoal filters.)

AIR CONDITIONING—the process of treating air so as to control simultaneously its temperature, humidity, cleanliness, and distribution to meet the requirements of the conditioned space.

AIR CONDITIONER, ROOM—encased assembly designed as a unit for mounting in a window, through a wall, or as a console.

AIR CONDITIONER, UNITARY—an evaporator, compressor, and condenser combination; designed in one or more assemblies, the separate parts designed to be assembled together.

ALTERNATING CURRENT—a common source of electrical energy, which reverses its direction of flow 50 to 60 times per second.

AMBIENT AIR—generally speaking, the air surrounding an object.

AMBIENT-COMPENSATED—designed so that varying temperatures of air at the control do not affect the control setting.

AMBIENT TEMPERATURE—the temperature of the air and/or other gases immediately surrounding a device.

AMPERAGE—the measure of current flow in an electrical circuit. Abbreviated: amp, A.

[1]The terms presented were assembled into a glossary by the Honeywell Corporation and are reproduced here with their permission.

ANEMOMETER—an instrument for measuring the velocity of a fluid.

ANTICIPATING CONTROL—one that by artificial means, is activated sooner than it would be without such means, to produce a smaller differential of the controlled property. Heat and cool anticipators are commonly used in thermostats.

ANTICIPATION—a method of reducing the operating differential of the system by adding a small resistive heater inside the thermostat to raise the internal temperature of the thermostat faster than the surrounding room temperature. This causes the thermostat to shut off the heating equipment and start the cooling equipment sooner than it would if affected only by the room temperature.

AQUASTAT—Honeywell trademark. A thermostat used to control water temperature.

ATOMIZE—to reduce a liquid into a multitude of tiny droplets or a fine spray.

AUXILIARY POTENTIOMETER—a potentiometer mounted on a modulating motor, which is used to control other modulating devices in response to the original motor's operation.

AUTOMATIC CONTROL—a system that reacts to a change or unbalance in the controlled condition by adjusting the variables, e.g., temperature and humidity, to restore the system to the desired balance.

AUXILIARY CONTACTS—a secondary set of electrical contacts mounted on a modulating motor whose operation coincides with the operation of the motor.

AVERAGING ELEMENT—a thermostat sensing element that responds to the average temperature sensed at many different points in a duct.

BILLING CHARGES (UTILITY)—a charge for the use of a unit of electricity. See Customer Charge, Demand Charge, and Energy Charge.

BILLING UNITS—units of electricity for which a charge is assessed. See Watts, Kilowatts, and Kilowatt Hours.

BIMETALLIC ELEMENT—one formed of two metals having different coefficients of thermal expansion such as are used in temperature indicating and controlling devices.

BLOWER—an air moving device; see Fan.

BOILER—a closed vessel in which a liquid is heated or vaporized.

BRITISH THERMAL UNIT, Btu—the amount of heat required to raise the temperature of a pound of water one degree Fahrenheit.

BULB—a thermostat sensing element, usually remote, which will respond to the temperature in the immediate vicinity of the bulb.

BUS BAR—a heavy, rigid metallic conductor which carries a large current and makes a common connection between several circuits. Bus bars are usually uninsulated and located where the electrical service enters a building; that is, in the main distribution cabinet.

BYPASS—a pipe or duct, usually controlled by valve or damper, for conveying a fluid around an element of a system.

cf—cubic feet. Standard measurement for natural gas.

cfm—cubic feet per minute; a measurement of volume flow. Metric equivalent is M^3/hr—cubic meters per hour.

CALORIE—the amount of heat required to raise the temperature of one gram of water one degree Celsius at one atmosphere of pressure.

CAPACITY—heating or cooling potential of a system, generally measured in tons or Btu's per hour.

CAPACITY CHARGE—see Demand Charge.

CAPILLARY TUBE—in refrigeration practice, a tube of small internal diameter used as a liquid refrigerant flow control or expansion device between high and low sides; also used to transmit pressure from the sensitive bulb of some temperature controls to the operating element.

CAPITAL INVESTMENT—an expenditure for an investment whose returns are expected to extend beyond one year.

CENTRAL FAN SYSTEM—an air conditioning system in which the air is processed at a central location outside the conditioned space and distributed by means of a fan and duct system.

CHANGEOVER—the process of switching an air conditioning system from heating to cooling, or vice versa.

CHILLED WATER SYSTEM—see Cooling System, Chilled Water.

CHILLER—the refrigeration system that cools the water in a chilled water system.

CLOSE-OFF—the maximum allowable pressure difference to which a valve may be subjected while fully closed.

COEFFICIENT OF PERFORMANCE—a term used to measure the efficiency of a heating system. It is defined as the heat output of a heat pump or electric elements, divided by the heating value of power consumed in watts at standard test conditions. Abbreviated: COP.

COIL—a cooling or heating element made of pipe or tubing.

COLD DECK—the cooling section of a mixed air zoning system.

COMBUSTION—the act or process of burning.

COMPRESSION—in mechanical refrigeration, process by which the pressure of the refrigerant is increased.

COMPRESSOR—the component of a mechanical refrigeration system which compresses the refrigerant vapor into a smaller volume, thereby raising the pressure of the refrigerant and consequently its boiling temperature.

COMPRESSOR, HERMETIC—see Compressor, Sealed Refrigerant.

COMPRESSOR, SEALED REFRIGERANT—(also called hermetic) a mechanical compressor combination consisting of a compressor and a motor, both of which are enclosed in the same housing, with no external shaft seals, the motor operating in the refrigerant atmosphere.

CONDENSATE—the liquid formed by condensation of a vapor. In steam heating, water condensed from steam; in air conditioning, water extracted from air, as by condensation on the cooling coil of a refrigeration machine.

CONDENSATION—the process of changing a vapor into liquid by the extraction of heat. Condensation of steam or vapor is effected in either steam condensers or in dehumidifying coils and the resulting water is called a condensate.

CONDENSER—arrangement of pipe or tubing in which a vapor is liquified by removal of heat.

CONDENSER, AIR-COOLED REFRIGERANT—a condenser cooled by natural or forced circulation of atmospheric air through it.

CONDENSER COIL—in mechanical refrigeration, a section of coiled tubing where gas refrigerant is cooled below its boiling point.

CONSUMPTION CHARGE—see Energy Charge.

CONTACTOR—electromagnetic switching device.

CONTROL—any device for regulation of a system or component in normal operation, manual or automatic. If automatic, it is responsive to changes of pressure, temperature, or other property whose magnitude is to be regulated.

CONTROL PANEL—an electrical cabinet that contains controls and/or indication devices.

CONTROL POINT—the value of the controlled variable that the controller operates to maintain.

CONTROLLED MEDIUM—the substance (usually air, water, or steam) whose characteristics (such as temperature, pressure, flow rate, volume, concentration, etc.) are being controlled.

CONTROLLED SPACE—the volume of the controlled medium; for example, a room in which the air temperature is being controlled.

CONTROLLED VARIABLE—that quantity or condition of a controlled medium which is measured and controlled; for example, temperature, pressure, flow rate, volume, concentration, etc.

CONVECTION—transfer of heat by movement of fluid.

COOLING SYSTEM, CHILLED WATER—a closed, circulating system in which a mechanical refrigeration system at a central location cools water which is then piped to various parts of the building.

COOLING SYSTEM, DIRECT EXPANSION (DX)—a cooling and dehumidification device which cools air or other fluids by the evaporation of mechanically compressed gas in an evaporative coil. A condensing coil then removes this transferred heat to a different space. See Refrigeration System, Mechanical.

COOLING SYSTEM, EVAPORATIVE—housed in a cabinet containing a pump, distribution tubes, water pads, and a blower. The pump supplies water to the distribution tubes which carry the water to pads on 3 sides of the cabinet. The blower draws outdoor air through the moist water pads. Some of the water in the pads absorbs heat from the air and evaporates. The air cannot be recirculated. This system works most effectively in relatively dry climates.

COOLING SYSTEM, MULTIPLE STAGE—a cooling system that increases its capacity by stages in response to cooling demand.

CRITICAL TIME OVERRIDE—an electromechanical device which bypasses the normal operation of a load management system during times that are extremely busy for building operation.

CRITICAL TIME PROGRAMMER—a clock that programs loads to bypass their normal cycling schedule during critical times.

CYCLE—a complete course of operation of working fluid back to a starting point, measured in thermodynamic terms (functions). Also used in general for any repeated process on any system.

CYCLE, REFRIGERATION—complete course of operation of refrigerant back to a starting point, evidenced by a repeated series of thermodynamic processes, or flow through a series of apparatus, or a repeated series of mechanical operations.

CYCLING RATE—the number of complete cycles that the system goes through in one hour. One complete cycle includes both on and off times.

DAMPER—an adjustable metal plate, louver, or set of louvers that controls airflow; especially through an air inlet, outlet, or duct.

DAMPER LINKAGE—linkage used to connect a motor to a damper, usually consisting of a pushrod, two crank arms, and two ball joints.

DAMPER, OPPOSED BLADE—alternate blades rotate in opposite directions. Provides an equal percentage flow characteristic— successive equal increments of rotation produce equal percentage increases in flow. Particularly useful for throttling applications where accurate control at low airflow is necessary.

DAMPER, PARALLEL BLADE—all blades rotate in the same direction. Provides a fairly linear airflow characteristic—the flow is nearly proportional to damper shaft rotation. Particularly useful in mixing applications where the sum of two airflows must be made constant.

DEADBAND—in HVAC, a temperature range in which neither heating nor cooling are turned on; in load management, a kilowatt range in which loads are neither shed nor restored.

DEGREE DAY—a unit, based upon temperature difference and time, used in estimating fuel consumption and specifying the nominal heating load of a building in winter. For any one day, when the mean temperature is less than 65°F, there exist as many degree days as there are Fahrenheit degrees difference in temperature between the mean temperature for the day and 65°F.

DEHUMIDIFICATION—the condensation of water vapor from air by cooling below the dew point or removal of water vapor from air by chemical or physical methods.

DEHUMIDIFIER—an air cooler or washer used for lowering the moisture content of the air passing through it. An absorption or adsorption device for removing moisture from the air.

DEHUMIDIFIER, WATER COIL—same as Cooling System, Chilled Water.

DEHYDRATION—the removal of water vapor from air by the use of absorbing or adsorbing materials. The removal of water from stored goods.

DELTA SERVICE—a 3 or 4 wire 3-phase wiring configuration commonly shown as "△."

DEMAND CHARGE—that part of an electric bill based on kW demand and the demand interval. Expressed in dollars per kilowatt. Demand charges offset construction and maintenance of a utility's need for a large generating capacity.

DEMAND CONTROL—a device that controls the kW demand level by shedding loads when the kW demand exceeds a predetermined set point.

DEMAND INTERVAL—the period of time on which kW demand is monitored and billed by a utility, usually 15 or 30 minutes long.

DEMAND READING—highest or maximum demand for electricity an individual customer registers in a given interval, for example, 15-minute interval. The metered demand reading sets the demand charge for the month.

DEVIATION—the difference between the set point and the value of the controlled variable at any instant.

DEW-POINT TEMPERATURE—the temperature at which moisture would begin to condense out of the air if the air should be cooled to that temperature. The temperature corresponding to saturation (100 percent relative humidity) for a given absolute humidity at constant pressure. The moisture content of the air establishes the dew-point temperature.

DIFFERENTIAL (of a control)—difference between the cut-out and cut-in points.

DIFFERENTIAL, INTERSTAGE—in a sequencing system, the amount of change in the controlled medium required to sequence from ON point of one stage to ON point of successive stage.

DIRECT CURRENT—a source of electrical power that flows in one direction only. Abbreviated: dc, V dc.

DIRECT EXPANSION (DX) SYSTEM—see Cooling System, Direct Expansion.

DISCHARGE AIR—conditioned air that is distributed to the controlled environment.

DISCRETE LOGIC—electronic circuitry composed of standard transistors, resistors, capacitors, etc., as compared to microprocessor circuits where the logic is condensed on a single chip (integrated circuit).

DRIER—device containing a dessicant, placed in the refrigerant circuit to collect and hold water in the system in excess of the amount that can be tolerated by the system refrigerated.

DRY AIR—air without water vapor; air only.

DRY BULB TEMPERATURE—the temperature of a gas or mixture of gases indicated by an accurate thermometer. Air temperature as read by an ordinary dry bulb thermometer.

DUCT—a passageway made of sheetmetal or other suitable material, not necessarily leak-tight, used for conveying air or other gas at low pressure.

DUTY CYCLING—energizing a load for part of a specified time period. Accomplished by a duty cycler.

ECONOMIZER—a system of dampers, temperature and humidity sensors, and motors that maximizes the use of outdoor air for cooling.

ECONOMIZER CONTROL—a system of ventilation control in which outdoor and return air dampers are controlled to maintain proper mixed air temperature for the most economical operation.

ELECTRICAL CIRCUIT—a power supply, a load, and a path for current flow are the minimum requirements for an electrical circuit.

ELECTROMECHANICAL—a term used to describe controls which contain both electrical and mechanical components.

ELECTRONIC AIR CLEANER—a device that produces a powerful electric field to ionize dirt and dust particles in the air. The particles are then collected on electrically charged plates.

ELEMENT, ELECTRIC HEATING—a unit assembly consisting of a resistor, insulated supports, and terminals for connecting the resistor to electric power.

ENERGY (CONSUMPTION) CHARGE—that part of an electric bill based on kWh consumption. Expressed in cents per kWh. Energy charge covers the cost of utility fuel, general operating costs, and part of the amortization of the utility's equipment.

ENERGY EFFICIENCY RATIO—a term used to measure the efficiency of an air conditioning system. It is defined as the number of Btu's removed, divided by the power consumed in watts at standard test conditions. Abbreviated: EER.

ENERGY MANAGEMENT—the control of the use of energy within a given facility or environment.

ENERGY MANAGEMENT SYSTEM—the set of devices, plans, or techniques which control the use of energy (electrical, gas, solar, etc.) within a given environment or facility.

ENTHALPY—the total heat content of the air. Includes sensible heat (air temperature) and latent heat (relative humidity).

ENVIRONMENTAL CONTROL SYSTEM—the process of controlling the environment by heating, cooling, humidifying, dehumidifying, or cleaning the air.

EVAPORATION—a change of state from liquid to vapor.

EVAPORATOR—the part of the refrigeration system where the refrigerant vaporizes (absorbs heat) to produce the cooling effect.

EVAPORATOR COIL— in mechanical refrigeration, a section of coiled tubing where liquid refrigerant absorbs heat and evaporates.

EXHAUST AIR—that air which is removed from the conditioned space by the ventilation system and discharged outdoors.

EXHAUST FAN—a fan that removes air from a space by exhausting or blowing it outdoors. Common in kitchens, bathrooms, and bars.

EXPANSION COIL—an evaporator constructed of pipe or tubing.

EXPANSION POINT—in a mechanical refrigeration system, a restriction or orifice which regulates the flow of refrigerant into the evaporator coil. May be in the form of a thermal expansion valve or a capillary tube.

EXPENSE—amortizing or writing off an investment in one year. A purchase.

FACE AND BYPASS DAMPER SYSTEM—a heating system in which the mixed airflow is divided into two duct sections, one through the coil (face) and the other around the coil (bypass). Dampers work in opposition (face damper closes while bypass damper opens, and vice versa) to regulate the amount of air that is heated. Used with either a steam or hot water coil. Demonstrates on-off control of coil and modulating control of airflow across coil.

FAIL SAFE—in Load Management, returning all controlled devices to conventional control in case of load management panel failure.

FAN COIL UNIT—a complete unit located in the room being conditioned consisting of a coil through which hot or cold water is circulated, a fan that circulates room air through the coil, a filter to remove lint and dust, a cabinet, a grille, and a control system. The boiler or chiller supplying the water is located centrally within the building.

FILTER—a device that removes solid material from liquids or gases such as air.

FLAME SAFEGUARD CONTROL—a control that provides a means for starting the burner in the proper sequence, proving that the burner flame is established, and supervising the flame during burner operation. May also provide a timing function (programmer) to sequence additional burner functions.

FLOW RATE—the rate at which fluids or gases will flow over a specified amount of time.

Units are gallons per minute (gpm) for water, pounds per hour for steam, and cubic feet per hour for gas.

FLUE—a passageway or conduit for conveying the combustion products (flue gases) to the outside.

FLUID—gas, vapor, or liquid.

FULL LOAD CURRENT—see Running Current.

FURNACE—that part of a boiler or warm air heating plant in which combustion takes place. Also a complete heating unit for transferring heat from the fuel being burned to the air supplied to a heating system.

HARD WIRING—permanent, line voltage (example: 120 or 240V ac) wiring.

HEADER—a manifold or supply pipe to which a number of branch pipes are connected.

HEAT EXCHANGER—a device specifically designed to transfer heat between two physically separated fluids.

HEAT OF FUSION—latent heat involved in changing between the solid and the liquid states.

HEAT OF VAPORIZATION—latent heat involved in the change between liquid and vapor states.

HEAT PUMP—mechanical refrigeration system with the added capability of reversing the refrigeration cycle to produce heating for the space. During heating, the functions of the evaporator and condenser are reversed.

HEAT RECLAIM—the process of reusing discharged heat from such sources as exhaust fans, condenser coils, and hot water drains to do useful work.

HEATER—see Heating System.

HEATING SYSTEM, DIRECT FIRED—a heating system in which combustion takes place in the air being blown into the building. The outdoor air temperature is increased by direct contact with the flame of the heater. Efficiency of this system is 100 percent.

HEATING SYSTEM, ELECTRIC HEATER—a heating system that consists of one or more stages of resistive heating elements installed in a duct or central furnace.

HEATING SYSTEM, DUCT HEATER—a heating system in which the heater is installed directly in the distribution duct of a central air conditioning or heating system. May be either electric, gas-fired, or oil-fired.

HEATING SYSTEM, HOT WATER COIL—a heating system in which hot water is supplied by a central hot water boiler. Hot water coils are used to heat mixed air.

HEATING SYSTEM, INDIRECT FIRED—a heating system in which combustion takes place in a boiler or furnace. The fuel is burned in a combustion chamber and the flue gases do not mix with the incoming air.

HEATING SYSTEM, RADIANT—a heating system in which only the heat radiated from panels is effective in providing the heating requirements. The term *radiant heating* is frequently used to include both panel and radiant heating.

HEATING SYSTEM, STEAM—a heating system in which heat is transferred from the boiler or other source of heat to the heating units by means of steam at, above, or below atmospheric pressure.

HEATING SYSTEM, WARM AIR—a warm air heating plant consisting of a heating unit (electric or fuel burning furnace) enclosed in a casing, from which the heated air is distributed to various rooms in a building through ducts.

HIGH LIMIT CONTROL—a device that normally monitors the condition of the controlled medium and interrupts system operation if the monitored condition becomes excessive.

HIGH SIDE—parts of the refrigerating system subjected to condenser pressure or higher; the system from the compression side of the compressor through the condenser to the expansion point of the evaporator.

HORSEPOWER—unit of power in foot-pound-second system; work done at the rate of 550 ft-lb per sec, or 33,000 ft-lb per min.

HOT DECK—the heating section of a multizone system.

HUMIDIFICATION—the process of increasing the water vapor content of the conditioned air.

HUMIDIFIER—a device to add moisture to the air.

HUMIDISTAT—a regulatory device, actuated by changes in humidity, used for the automatic control of relative humidity.

HUMIDITY—water vapor within a given space.

HUMIDITY RATIO—see Specific Humidity.

HUNTING—an undesirable condition where a controller is unable to stabilize the state of the controlled medium and cycles rapidly.

HVAC—Heating, Ventilating, and Air Conditioning.

HYDRONIC SYSTEM—a heating and/or cooling system that uses a liquid (usually hot or cold water) as the medium for heat transfer.

HYDRONICS—the science of heating and cooling with liquids.

HYDROMETER—an instrument that, by the extent of its submergence, indicates the specific gravity of the liquid in which it floats.

HYGROMETER—an instrument responsive to humidity conditions (usually relative humidity) of the atmosphere.

HYGROSTAT—same as Humidistat.

"IN" CONTACTS—those relay contacts which complete circuits when the relay armature is energized. Also referred to as Normally Open Contacts.

INDUCTIVE LOADS—loads whose voltage and current are out-of-phase. True power consumption for inductive loads is calculated by multiplying its voltage, current, and the power factor of the load.

INFILTRATION—in air conditioning, the natural leakage of fresh outdoor air into a building.

INRUSH CURRENT—the current that flows the instant after the switch controlling current flow to a load is closed. Also called Locked Rotor Current.

INTERSTAGE DIFFERENTIAL—in a multistage HVAC system, the change in temperature at the thermostat needed to turn additional heating or cooling equipment on.

ISOTHERMAL PROCESS—a process in which there is no change in dry bulb temperature.

KILOWATT—1000 watts. Abbreviated: kW.

KILOWATT-HOUR—a measure of electrical energy consumption. 1000 watts being consumed per hour. Abbreviated: kWh.

kW DEMAND—the maximum rate of electrical power usage for a 15- or 30-minute interval in a commercial building for each billing period. A utility meter records this maximum rate, and customers are billed for this peak rate usually once per month.

kWh CONSUMPTION—the amount of electrical energy used over a period of time; the number of kWh used per month. Often called Consumption.

LAG—a delay in the effect of a changed condition at one point in the system, on some other condition to which it is related. Also, the delay in action of the sensing element of a control, due to the time required for the sensing element to reach equilibrium with the property being controlled; i.e., temperature lag, flow lag, etc.

LATENT HEAT—the amount of heat necessary to change a quantity of water to water vapor without changing either temperature or pressure. When water is vaporized and passes into the air, the latent heat of vaporization passes into the air along with the vapor. Likewise, latent heat is removed when water vapor is condensed.

LIGHT EMITTING DIODE—a low current and voltage light used as an indicator. Abbreviated: LED.

LIMIT—control applied in the line or low voltage control circuit to break the circuit if conditions move outside a preset range. In a motor, a switch that cuts off power to the motor windings when the motor reaches its full open position.

LIMIT CONTROL—a temperature, pressure, humidity, dew-point, or other control that is used as an override to prevent undesirable or unsafe conditions in a controlled system.

LIMIT SHUTDOWN—a condition in which the system has been stopped because the value of the temperature or pressure has exceeded a pre-established limit.

LINE VOLTAGE—in the control industry, the normal electric supply voltages, which are usually 120 or 240 volts.

LIQUID LINE—the tube or pipe carrying the refrigerant liquid from the condenser or receiver of a refrigerating system to the evaporator or other pressure-reducing device.

LOAD—that part of an electrical circuit in which useful work is performed. In a heating or cooling system, the heat transfer that the system will be called upon to provide. Any equipment that can be connected to a load management system.

LOAD FACTOR—a comparison of kilowatt-hours of electricity consumed to the peak rate at which power was consumed. Load factor is always a number between zero and one and is expressed as the kilowatt-hours consumed over the specified period divided by the product of the kilowatt peak demand registered times the number of hours in the period.

LOAD MANAGEMENT—the control of electrical loads to reduce kW demand and kWh consumption.

LOAD MANAGEMENT SYSTEM—the set of devices that, when installed, effectively reduces kW and kWh consumption.

LOAD OVERRIDE—see Limit Control.

LOAD PROGRAMMER—any device that turns loads on and off on a real time, time interval, or kW demand basis.

LOAD RELAY—a relay that directly switches a load.

LOAD SCHEDULING—see Time-of-Day Programming.

LOCKED ROTOR CURRENT—see Inrush Current.

LOUVER—an assembly of sloping vanes intended to permit air ventilation to pass through and to inhibit the transfer of water droplets.

LOW LIMIT CONTROL—a device that normally monitors the condition of the controlled medium and interrupts system operation if the monitored condition drops below the desired minimum value.

LOW SIDE—the refrigerating system from the expansion point to the point where the refrigerant vapor is compressed; where the system is at or below evaporator pressure.

LOW VOLTAGE—in the control industry, a power supply of 25 volts or less.

MAIN—a pipe or duct for distributing to, or collecting from, various branches.

MAKEUP AIR—outdoor air that is brought into a building to compensate for air removed by exhaust fans or other methods.

MAKEUP AIR SYSTEM—a system of replacing exhausted air with fresh outdoor air that is then heated or cooled. The system uses a blower to take in air, an indirect heater (boiler or furnace) or a direct heater located in the air duct, and a set of controls to regulate the air temperature.

MAIN DISTRIBUTION CABINET—the enclosure in a facility that houses utility metering equipment, the main electrical disconnect switch, and the branch circuit overcurrent protection. Electrical service is distributed throughout the building from this cabinet.

MANIFOLD—portion of a main in which several branches are close together. Also, single piece in which there are several fluid paths. Also called Header.

MANOMETER—an instrument for measuring pressures; essentially a U-tube partially filled with a liquid, usually water, mercury, or a light oil, so constructed that the amount of displacement of the liquid indicates the pressure being exerted on the instrument.

MECHANICAL REFRIGERATION SYSTEM—a cooling system consisting basically of a refrigerant, compressor, evaporator coil, and condenser coil. In a basic cycle, the refrigerant is compressed, liquefied, and cooled below its boiling point. It then enters the evaporator coil where it expands and boils, absorbing heat from its surroundings. It is then compressed again and a new cycle begins.

MEDIUM, HEATING—a solid or fluid, such as water, steam, air or flue gas, used to convey heat from a boiler, furnace or other heat source, and to deliver it, directly or through a suitable heating device, to a substance or space being heated.

MICROPROCESSOR—a small computer used in load management to analyze energy demand and consumption such that loads are turned on and off according to a predetermined program.

MINIMUM FRESH AIR REQUIREMENTS—the amount of fresh, outdoor air needed to replenish interior space air to satisfy local legal codes.

MINIMUM ON-TIME—the shortest period of time that a load can be energized when it is being duty cycled.

MIXED AIR—that air in a ventilation system that is composed of return air and outdoor air before it has been conditioned.

MIXING BOX—a container, located at the room being conditioned, in which hot and cold air is mixed as required to maintain the desired room temperature.

MODULATING CONTROL—a mode of automatic control in which the action of the final control element is proportional to the deviation, from set point, of the controlled medium.

MODULATING MOTOR—an electric motor, used to drive a damper or valve, which can position the damper or valve anywhere between fully open or fully closed in proportion to deviation of the controlled medium.

MODULATING RANGE—see Proportional Band.

MODUTROL MOTOR—Honeywell trademark. A line of two-position and modulating motors used to control dampers or valves.

MOISTURE CONTENT—amount of water vapor in a given amount of air, usually expressed in grains of moisture per lb of dry air. (7000 grains are equal to 1 lb.)

MORNING PICKUP—see Morning Warmup.

MORNING WARMUP—a control system that keeps outside air dampers closed after night setback until the desired space temperature is achieved.

MULTIPLE STAGE SYSTEM—see Cooling System, Multiple Stage.

MULTISTAGE THERMOSTAT—a temperature control that sequences two or more switches in response to the amount of heating or cooling demand.

MODULATING—tending to adjust by increments and decrements.

MULTIZONE UNIT—see Zoning, Mixed Air.

MULTIZONE SYSTEM—centralized HVAC system that controls several zones with each zone having a thermostat.

N.C.—normally closed contacts of a relay. Contacts are close-circuited when the relay is deenergized.

N.O.—normally open contacts of a relay. Contacts are open-circuited when relay is deenergized.

NIGHT SETBACK—the ability to reduce heating expense during unoccupied hours by lowering temperature, closing outside air dampers, and intermittently operating blowers.

OFFSET—a sustained deviation between the actual control point and the set point under stable operating conditions.

ON-OFF CONTROL—a simple control system, consisting basically of a switch, in which the device being controlled is either fully on or fully off and no intermediate positions are available.

ORIFICE—an opening or construction in a passage to regulate the flow of a fluid.

"OUT" CONTACTS—those relay contacts which complete circuits when the relay coil is deenergized. Also referred to as Normally Closed Contacts.

OUTDOOR AIR—air that is brought into the ventilation system from outside the building and, therefore, not previously circulated through the system

OVERRIDE—a manual or automatic action taken to bypass the normal operation of a device or system.

PACKAGED MULTIZONE UNITS—a packaged heating, ventilating, air conditioning unit that simultaneously maintains separate temperatures on a hot and cold deck. Each zone within the building then mixes air from these two decks to maintain space temperature.

PACKAGED SYSTEM—a complete set of components and controls factory-assembled for ease of installation. A packaged system may perform one or more of the air conditioning functions.

PACKAGED TERMINAL AIR CONDITIONER (PTAC)—an electrical heating and cooling system often found in the guest rooms of hotel/motels. The unit is installed through the wall of the room.

PANEL POWER FAILURE RELAY—a relay that sheds noncritical loads during a power failure to the load management panel. Used to avoid setting a demand peak when power is lost to the load management panel.

PAYBACK—total investment divided by one year's savings. This gives the number of years it will take to recover the investment with no regard to interest rates or taxes.

PEAK DEMAND—the greatest amount of kilowatts needed during a demand interval.

PHASE—an electrical term used to describe the number of distinct harmonic waves in alternating current electrical services. Residential service is single-phase; commercial facilities are usually three-phase.

PILOT DUTY RELAY—a relay used for switching loads such as another relay or solenoid valve coils. The pilot duty relay contacts are located in a second control circuit. Pilot duty relays are rated in volt-amperes (VA).

PLENUM CHAMBER—an air compartment connected to one or more distributing ducts.

PNEUMATIC—operated by air pressure.

POTENTIAL TRANSFORMER—a voltage transformer. The voltage supplied to a primary coil induces a voltage in a secondary coil according to the ratio of the wire windings in each of the coils.

POTENTIOMETER—an electromechanical device having a terminal connected to each end of the resistive element, and third terminal connected to the wiper contact. The electrical input is divided as the contact moves over the element, thus making it possible to mechanically change the resistance.

POWER—in electricity, the watt. A time rate measurement for the use of electrical energy. Joules per second.

POWER FACTOR—a ratio, sometimes expressed as a percent of actual power (watts) in an ac circuit to apparent power (volt-amperes). A measure of power loss in an inductive circuit. When the power factor is less than 0.8, the utility may impose a penalty, as prescribed in the utility rate structure.

POWER FACTOR CHARGE—a utility charge for "poor" power factor. It is more expensive to provide power to a facility with a poor power factor (usually less than 0.8).

POWER FACTOR CORRECTION—installing capacitors on the utility service's supply line to improve the power factor of the building.

POWER SUPPLY—the voltage and current source for an electrical circuit. A battery, a utility service, and a transformer are power supplies.

PREHEAT—a process of raising the temperature of outdoor air before incorporating it into the rest of the ventilating system. Used when large amounts of very cold outdoor air must be used.

PRESSURE—the normal force exerted by a homogeneous liquid or gas, per unit of area, on the wall of its container.

PRESSURE, ABSOLUTE—sum of gauge pressure and atmospheric pressure. Absolute pressure can be zero only in a perfect vacuum.

PRESSURE, ATMOSPHERIC—the pressure exerted in every direction at any given point by the weight of the atmosphere. It is the pressure indicated by a barometer. Standard Atmospheric Pressure or Standard Atmosphere is the pressure of 76 cm of mercury having a density of 13.5951 grams per cm³, under standard gravity of 980.665 cm per sec. It is equivalent to 14.696 psi or 29.921 inches of mercury at 32°F.

PRESSURE, SUCTION—the refrigerant pressure as measured at the inlet of a compressor in a direct expansion refrigeration system.

PRESSURE CONTROLS—Pressure controls are used as limit protectors in the cooling system. They establish pressure control limits to protect the system from extremes in refrigeration suction and discharge line pressures. If the pressure deviates from normal, the pressure control breaks the circuit to the compressor until the pressure returns to normal.

Pressure controls have automatic or manual reset, depending upon the construction of the equipment and preference of the manufacturer.

PRESSURE DROP—the difference between the upstream pressure and the downstream pressure of a fluid passing through a valve. Symbol: h.

PRESSURE, GAUGE—pressure measured above atmospheric pressure; indicated by a manometer.

PRESSURE, HEAD—operating pressure measured in the discharge line at a compressor outlet.

PRESSURE REGULATOR—automatic valve between the evaporator outlet and compressor inlet that is responsive to pressure or temperature; it functions to throttle the vapor flow when necessary to prevent the evaporator pressure from falling below a preset level.

PRIMARY CONTROL—a device that directly or indirectly controls the control agent in response to needs indicated by the controller. Typically, a motor, valve, relay, etc.

PROPORTIONAL BAND—the range of values of a proportional positioning controller through which the controlled variable must pass to move the final control element through its full operating range. Commonly used equivalents are "throttling range" and "modulating range."

PROPORTIONAL CONTROL—see Modulating Control.

PSYCHROMETER—an instrument with wet and dry bulb thermometers, for measuring the amount of moisture in the air. See Wet Bulb Temperature.

RADIATION, THERMAL (HEAT)—the transmission of energy by means of electromagnetic waves of very long wavelength. Radiant energy of any wavelength may, when absorbed, become thermal energy and result in an increase in the temperature of the absorbing body.

RADIATOR—a heating unit exposed to view within the room or space to be heated. A radiator transfers heat by radiation to objects within visible range, and by conduction to the surrounding air which, in turn, is circulated by natural convection; a so-called radiator is also a convector, but the term *radiator* has been established by long usage.

RAPID CYCLING—see Short Cycling.

RECEIVER—storage chamber for liquid refrigerant in a mechanical refrigeration system; often the bottom part of the condenser.

RECIRCULATED AIR—return air passed through the conditioner before being again supplied to the conditioned space.

REFRIGERANT—a substance with a large latent heat of vaporization and low boiling point that produces a refrigerating effect by absorbing heat while expanding or vaporizing (boiling).

REFRIGERATING SYSTEM, ABSORPTION—a refrigerating system in which the refrigerated gas evolved in the evaporator is taken up in an absorber and released in a generator upon the application of heat.

REFRIGERATING SYSTEM, CENTRAL PLANT—a system with two or more low sides connected to a single, central high side; a multiple system.

REFRIGERATING SYSTEM, CHILLED WATER—an indirect refrigerating system employing water as the circulating liquid.

REFRIGERATING SYSTEM, COMPRESSION—a refrigerating system in which the pressure-imposing element is mechanically operated.

REFRIGERATING SYSTEM, DIRECT-EXPANSION—a refrigerating system in which the evaporator is in direct contact with the refrigerated material or space or is located in air circulating passages communicating with such spaces.

REFRIGERATING SYSTEM, MECHANICAL—system where the evaporator coil produces cooling by absorbing heat from the surrounding air, raising the refrigerant to its boiling point and causing it to vaporize. The superheated vapor flows through the condenser, which condenses it into a liquid and gives off heat picked up in the evaporator coil. Then the liquid flows to the expansion point, where it expands (lowering its temperature and pressure) to start the cooling cycle again.

REFRIGERATING SYSTEM, MULTIPLE—a refrigerating system using the direct method in which the refrigerant is delivered to two or more evaporators in separate rooms or refrigerators.

REFRIGERATING SYSTEM, SINGLE-PACKAGE—a complete factory-made and factory-tested refrigerating system in a suitable frame or enclosure, which is fabricated and shipped in one or more sections and in which no refrigerant-containing parts are connected in the field.

REFRIGERATION SYSTEM—combination of interconnected refrigerant-containing devices in which the refrigerant is circulated for the purpose of extracting heat to produce cooling.

REFRIGERATOR—a container and means for cooling it, such as a domestic refrigerator, or a large container such as a storage refrigerator, service refrigerator, etc.

REFRIGERATOR, ELECTRIC—a completely self-contained unit consisting of an insulated cabinet, evaporator, condensor, and an electric compressor.

REFRIGERATOR, GAS—a refrigerator motivated by thermal energy of burning gas.

REGISTER—a combination grille and damper assembly covering an air opening.

REGULATOR LAG—the lapse of time, usually 9 to 12 months, between a petition for rate increase filed by a utility and the formal action on the petition by the public service commission.

REHEAT—the process of adding heat to air to maintain the correct temperature after it has previously been cooled to some specified dew point to control humidity.

RELATIVE HUMIDITY—the ratio of the existing vapor pressure of the water in the air to the vapor pressure of water in saturated air at the same dry bulb temperature.

RELAY—an electromechanical switch that opens or closes contacts in response to some controlled action. Relay contacts are normally open (N.O.) and normally closed (N.C.).

RELAY, MAGNETIC—solenoid-operated relay or contactor; a switching relay that utilizes an electromagnet (solenoid) and an armature to provide the switching force.

RELAY, THERMAL—a switching relay in which a small heater warms a bimetal element that bends to provide the switching force.

REMOTE TEMPERATURE SET POINT—ability to set a temperature control point for a space from outside the space. Often used in public areas.

RESET—a process of automatically adjusting the control point of a given controller to compensate for changes in the outdoor temperature. The hot deck control point is normally reset upward as the outdoor temperature drops. The cold deck control point is normally reset downward as the outdoor temperature increases.

RESET RATIO—the ratio of change in outdoor temperature to the change in control point temperature. For example, a 2:1 reset ratio means that the control point will increase 1 degree for every 2 degrees change in outdoor temperature.

RESISTANCE—the opposition that limits the amount of current that can be produced by an applied voltage in an electrical circuit. Measured in ohms, abbreviated with the Greek letter omega (Ω).

RESISTIVE LOADS—electrical loads whose power factor is one. Usually contain heating elements.

RESTORE—to energize a load that has been shed.

RETURN AIR—air that is drawn back into the ventilation system from the controlled space.

ROOFTOP UNIT—HVAC system placed on a roof and connected to ducts that supply conditioned air to the area below it.

ROTATING LOADS—alternately shedding and restoring loads assigned to a specific channel so that they will not be shed continuously. Also called Alternating Channels.

RUNNING CURRENT—the current that flows through a load after inrush current. Usually called Full Load Current.

SEASONAL PEAK—the maximum demand placed on the utility's capacity resulting from seasonal factors. Some utilities have summer peaks, some winter peaks, some both.

SENSIBLE HEAT—that heat which changes the temperature of the air without a change in moisture content. Changes in dry bulb thermometer readings are indicative of changes in sensible heat.

SENSING ELEMENT—the first system element or group of elements. The sensing element performs the initial measurement operation.

SENSOR—a sensing element.

SEQUENCER—a mechanical or electrical device that may be set to initiate a series of events and to make the events follow in sequence.

SEQUENCING CONTROL—a control that energizes successive stages of heating or cooling equipment as its sensor detects the need for increased heating or cooling capacity. May be electronic or electromechanical. See Sequencer.

SET POINT—the value on the controller scale at which the controller indicator is set.

SHED—to deenergize a load in order to maintain a kW demand set point.

SHIELDED CABLE—special cable used with equipment that generates a low voltage output. Used to minimize the effects of frequency "noise" on the output signal.

SHORT CYCLING—unit runs and then stops at short intervals; generally, this excessive cycling rate is hard on the system equipment.

SINGLE-ZONE SYSTEM—HVAC system controlled by one thermostat.

SPACE THERMOSTAT—a thermostat whose sensor is located in the controlled space.

STACK—a chimney; an exhaust pipe. One or several flues or pipes arranged to let the products of combustion escape to the outside.

STAGE DIFFERENTIAL—change in temperature at the thermostat needed to turn heating or cooling equipment off once it is turned on.

STAGING INTERVAL—the minimum time period for shedding or restoring two sequential loads.

STARTER—basic contactor with motor overloads, etc., added—A motor starter is an adaptation of the basic contactor which includes overload relays. Starters for large motors may include reactors, step resistors, disconnects, or other features required in a more sophisticated starter package.

STEP CONTROLLER—see Sequencer.

SUBCOOLING—liquid temperature and pressures are directly related; subcooling is cooling the liquid line below the saturation temperature corresponding to the pressure. Subcooling is measured by comparing the temperature of the liquid line to what the liquid temperature would ordinarily be at the measured discharge pressure.

SUBLIMATION—change of state directly from solid to gas without appearance of liquid.

SUCTION INLET—port through which gas enters the compressor.

SUCTION LINE—the tube or pipe that carries the refrigerated vapor from the evaporator to the compressor inlet.

SUPERHEAT—vapor temperature and pressure are directly related; superheating is raising the temperature of the refrigerant vapor above the saturation temperature corresponding to the pressure. Superheat is measured by comparing the temperature of the vapor to what the vapor temperature would ordinarily be at the measured pressure.

SWITCHING RELAYS—relays are devices that operate by a variation in the conditions of one electrical circuit to affect the operation of devices in the same or another circuit. General purpose switching relays are used to increase switching capability and isolate electrical circuits, such as in systems where the heating and cooling equipment have separate power supplies, and provide electrical interlocks within the system.

SYSTEM, FORCED-CIRCULATION—a heating, air conditioning, or refrigerating system in which the heating or cooling fluid circulation is effected by a fan or pump.

SYSTEM, GRAVITY CIRCULATION—a heating or refrigerating system in which the heating or cooling fluid circulation is effected by the motive head due to difference in densities of cooler and warmer fluids in the two sides of the system.

TEMPERATURE—the thermal state of matter with reference to its tendency to communicate heat to matter in contact with it. If no heat flows upon contact, there is no difference in temperature.

TERMINAL REHEAT SYSTEM—centralized blower and cooling system that supplies cool air to multiple zones. Each zone contains a hot water coil or electric heater that reheats the cooled supply air as determined by the zone thermostat.

THERM—measurement used by gas utilities for billing purposes. 1 Therm = 100 cf of gas = 100,000 Btu's.

THERMISTOR—semiconductor material that responds to temperature changes by changing its resistance.

THERMOCOUPLE—device for measuring temperature utilizing the fact that an electromotive force is generated whenever two dissimilar metals in an electric circuit are at different temperature levels.

THERMOSTAT—the thermostat serves as the basic controller for the HVAC system. It senses space temperature and signals a requirement for heating or cooling to maintain the temperature set point.

THROTTLING RANGE—the range of values of a proportional controller through which the controlled variable must pass to drive the final control element through its full operating range. Also called Proportional Band. In a thermostat, the temperature change required to drive the potentiometer wiper from one end to the other, typically 3 to 5°F.

TIGHT SHUTOFF—virtually no flow or leakage through the valve in its closed position.

TIME CLOCK—a mechanical or electrical device used to monitor the actual time of day.

TON OF REFRIGERATION—refrigerating effect equal to 12,000 Btu per hour, or the amount of heat required to melt one ton of ice.

TOTAL HEAT—the sum of the sensible and latent heats. Changes in wet bulb thermometer readings are indicative of changes in total heat. For convenience total heat is measured from 0°F.

TRANSFORMER—the system power supply—a transformer is an inductive stationary device that transfers electrical energy from one circuit to another. The transformer has two windings, primary and secondary. A changing voltage applied to one of these, usually the primary, induces a current to flow in the other winding. A coupling transformer transfers energy at the same voltage; a step-down transformer transfers energy at a lower voltage, and a step up transformer transfers energy at a higher voltage.

TWO-POSITION CONTROL—see On-Off Control.

UNIT COOLER—a direct-cooling, factory-made, encased assembly including a cooling element, fan and motor (usually), and directional outlet.

UNIT HEATER—a direct-heating, factory-made, encased assembly including a heating element, fan and motor, and directional outlet.

UNITARY SYSTEM—a room unit that performs part or all of the air conditioning functions. It may or may not be used with a central fan system.

UNLOADER—a device on or in a compressor for equalizing the high and low side pressures for a brief period during starting, in order to decrease the starting load on the motor; also a device for controlling compressor capacity by rendering one or more cylinders ineffective.

UTILITY SERVICE—the utility company. Also, the amount and configuration of voltage supplied by a utility company. There are four main types of commercial utility services: 208 volts ac wye, 480 volts ac wye, 240 volts ac delta, and 480 volts ac delta.

UTILITY TRANSFORMER—primary and secondary coils of wire that reduce (step down) the utility supply voltage for use within a facility.

VALVE—a device in a pipe or tube that diverts, mixes, or stops fluid flow by means of a flap, lid, plug, etc., acting to open and/or block passages.

VALVE, AUTOMATIC CONTROL—a valve combines a valve body and a valve actuator or motor. A signal from some remote point can energize the actuator either to open or close the valve, or to proportion the rate of flow through the valve.

VALVE BODY—the portion of the valve through which the medium flows.

VALVE, CHECK—valve that allows fluid to flow in one direction only.

VALVE, EXPANSION—valve that controls the refrigerant flow to the evaporator.

VALVE, MODULATING—a valve that can be positioned anywhere between fully on and fully off to proportion the rate of flow in response to a modulating controller (see Modulating Control).

VALVE, ON-OFF—see Valve, Two-Position.

VALVE, SOLENOID—an automatic, two-position valve that is either opened or closed by the action of an electrically excited coiled wire magnet (electromagnet) upon a bar of steel attached to the valve disc.

VALVE, THERMOSTATIC RADIATOR—a device that combines a temperature-sensitive power unit with a valve. Mounts in the hot water or steam supply pipe adjacent to a radiator. May also be applied to baseboard and similar types of heaters. Requires no electricity.

VALVE, TWO-POSITION—a valve that is either fully on or fully off with no positions between. Also called an On-Off Valve.

VARIABLE AIR VOLUME SYSTEM—centralized HVAC system that supplies conditioned air to zones where a regulator and a thermostat determine the volume of air delivered to the space.

VELOCITY—a vector quantity that denotes at once the time rate and the direction of a linear motion.

VELOCITY, OUTLET—the average discharge velocity of primary air being discharged from the outlet, normally measured in the plane of the opening.

VELOCITY, ROOM—the average sustained residual air velocity level in the occupied zone of the conditioned space; e.g., 65, 50, 35 fpm.

VELOCITY, TERMINAL—the highest sustained airstream velocity existing in the mixed air path at the end of the throw.

VENTILATION—the process of supplying or removing air, by natural or mechanical means, to or from any space. Such air may or may not have been conditioned.

VOLTAGE—the electromotive force in an electrical circuit. The difference in potential between two unlike charges in an electrical circuit is its voltage. Abbreviated: V.

WATT—a measure of electric power. Equal to a current flow of one ampere under one volt of pressure. One joule per second. About 1/746 horsepower. Abbreviated: W, w.

WATT TRANSDUCER—a device that converts a current signal into a proportional millivolt signal. Used to interface between current transformers and a load management panel.

WET BULB TEMPERATURE—an air temperature measurement that can be used to determine the relative humidity of air. The term is derived from the fact that a thermometer bulb is encased in a wick soaked with water.

WYE SERVICE—a four-wire, three-phase wiring configuration. Commonly shown as ''Y.''

ZONING—the practice of dividing a building into small sections for heating and cooling control. Each section is selected so that one thermostat can be used to determine its requirements.

ZONING, MIXED AIR—hot and cold air are mixed in just the right proportions to maintain the desired zone temperature. The air is channeled into a heating section (hot deck) and cooling section (cold deck) and then mixed.

ZONING, MULTIPLE UNIT—a separate heating-cooling unit is used for each zone.

ZONING, VOLUME—has just one central heating-cooling unit with ducts leading to each zone.

Appendix A Design Problems

A series of design problems is presented in this appendix. The problems vary in complexity and may be assigned as homework or as semester projects. As with any design problem there may be more than one solution.

DESIGN PROBLEM A

A unit heater with a 230-volt fan motor hangs in a warehouse to provide heat. A line voltage thermostat is to energize the fan on a call for heat. A thermostatic switch mounted on the hot water supply line is to prevent the fan motor from operating if the supply water temperature is less than 120°F.

Draw the electrical wiring diagram that will satisfy these requirements. See Figure A-1.

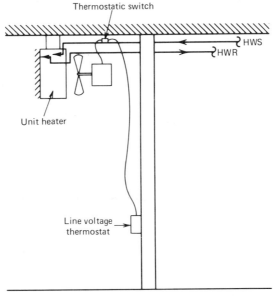

FIGURE A-1

DESIGN PROBLEM B

A factory area is to be ventilated. During the summer makeup air temperature is uncontrolled whereas in the winter an oil-fired makeup air unit heats the air to 65°F prior to delivering it to the conditioned space. The system is to be in operation 5 days per week from 8 A.M. to 4 P.M. and will be automatically controlled. A planview of the factory is shown in Figure A-2.

Draw a control diagram and describe the sequence of operation for this system.

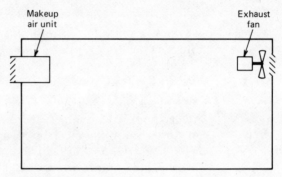

FIGURE A-2

DESIGN PROBLEM C

A clothing store is to be located in a small shopping mall. It will have a single-zone roof-top gas-fired heating, electric cooling forced air system. An automatic changeover control based on the outdoor temperature is to be incorporated in this system along with an economizer cycle for free cooling during the spring and fall.

Draw a control diagram and describe the sequence of operation for this system. See Figure A-3.

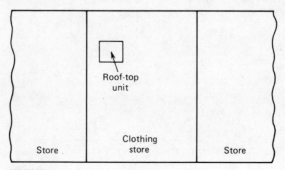

FIGURE A-3

DESIGN PROBLEM D

A motel with 50 guest rooms is to have a two-pipe fan coil system with a manual changeover. A chiller system will deliver water at a constant 45°F temperature. It employs an air-cooled condenser with head pressure control consisting of dampers over the face of the condenser coil responsive to high side pressure. The heating system is an oil-fired hot water generator designed to deliver hot water at a temperature determined by a master-submaster control system with a 1:1.5 reset ratio. At 60°F outdoor temperature the boiler water is at 100°F.

Draw a control diagram for the cooling system, heating system, and fan coil unit. Describe the sequence of operation of each system.

DESIGN PROBLEM E

The operating room of a hospital is to be kept at a constant temperature of 78°F and a relative humidity of 50%. A 100% outdoor air system is to be used and a slight positive pressure is to be maintained in the room relative to the surrounding corridors. The main hospital chilled water system and steam supply will be used to supply the heating and cooling coils employed in this ducted system.

Draw a control diagram and describe the sequence of operation of this system. See Figure A-4.

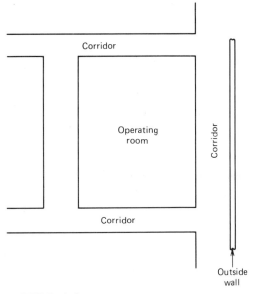

FIGURE A-4

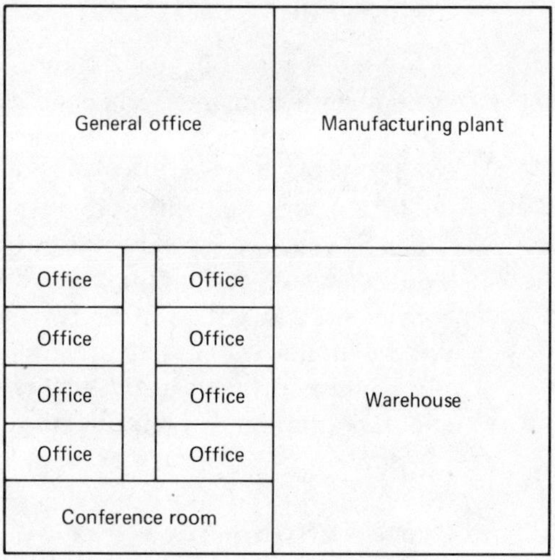

FIGURE A-5

DESIGN PROBLEM F

The manufacturing plant shown in Figure A-5 is located in a temperate zone and requires heating and cooling. You are the HVAC engineer and must determine the type of system you will recommend to the owner as well as the control system required. Your recommendation will take the following form:

1. A written report describing the type of system, how it is controlled, and the reason for its selection.

2. A drawing of the building with a single-line representation of the duct system and/or piping system as well as placement of the major equipment such as the boiler, cooling tower, and so forth. Also show the location of thermostats and other controls.

3. A control drawing, similar to those in the text, for each major piece of equipment, that is, multizone units, VAV units, mixing boxes, and so forth. A written sequence of operation for each system is also required.

Appendix B Electrical Wiring Diagram Exercises

This appendix contains a series of electrical wiring diagram exercises designed to illustrate how electrical components are interconnected. The student is asked to draw, complete, or analyze these diagrams. There may be more than one correct solution to a problem.

DIAGRAM 1   DRAW A WIRING DIAGRAM

In the space below complete the schematic wiring diagram (Figure B-1) with the following components: 120-volts ac fused power supply, three light bulbs, two SPST switches, and one DPST relay with a line voltage coil. The system must be such as to control two of the light bulbs with the relay and one of the bulbs with a SPST switch. The relay coil is to be controlled with the other SPST switch.

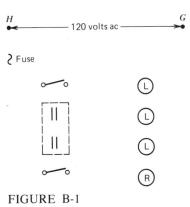

FIGURE B-1

DIAGRAM 2

Draw a wiring schematic diagram for a fan coil unit with the following components: a line voltage thermostat, motorized zone valve, "on-off" switch, and fan motor. Assume that all loads are rated at 115 volts. The "on-off" switch energizes the fan motor, permitting it to run constantly. It also provides power to the thermostat that controls the motorized zone valve.

DIAGRAM 3

A humidifier has a small 115-volt motor that causes a belt to rotate through a reservoir of water and into a stream of hot air on the outlet side of a warm air furnace. A SPST humidistat senses the return air humidity and controls the humidifier motor. A sail switch is installed downstream of the furnace to insure that the humidifier operates only when the furnace fan is in operation.

Draw a schematic wiring diagram of how such a humidification system could be wired.

DIAGRAM 4

A water-cooled air conditioning unit has a 230/1/60 compressor and evaporator fan motor. Both the fan motor and compressor motor are controlled by a single DPST contactor with a 24-volt coil. A 230/24-volt transformer provides low voltage power to a remote cooling thermostat. The compressor is protected by line voltage controls sensing low pressure and high pressure in the refrigeration system.

Draw a wiring schematic diagram depicting this system using the following symbols.

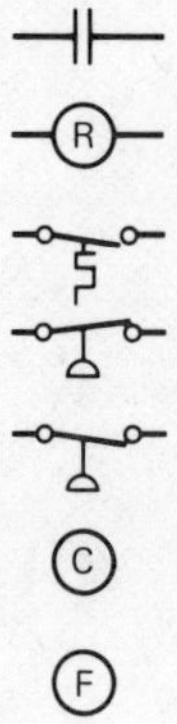

R   NO contact

R   Contactor coil

Cooling thermostat

High pressure cutout

Low pressure cutout

C   Compressor

F   Fan motor

DIAGRAM 5

An electric duct heater is made up of two 10-kilowatt heating elements rated at 230/1/60. A DPST contactor is required for each heater. The heaters are intended to respond to a two-stage low voltage thermostat with a 1°F temperature differential between stages. Each heater has a fusible link as final protection, a manual reset line voltage bimetal over-temperature protection switch, and an automatic reset, low voltage bimetal over-temperature protection switch. A 230/24-volt transformer provides low voltage power. A sail switch, sensing airflow in the duct, is wired in series with the thermostat. Using the following symbols, draw a wiring schematic that will meet these requirements.

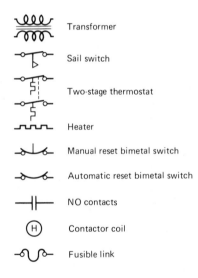

Transformer

Sail switch

Two-stage thermostat

Heater

Manual reset bimetal switch

Automatic reset bimetal switch

NO contacts

Contactor coil

Fusible link

DIAGRAM 6

A water pump is to be driven by a 208/3/60 electric motor. A line voltage cooling thermostat in series with the coil of a three-pole contactor is to control the operation of the pump. Draw the wiring schematic for such an arrangement.

DIAGRAM 7

A fan motor with a rated voltage of 208/3/60 is to be controlled by a motor starter. The coil of the starter has a 24-volt rating and is in series with a low voltage thermostat used for heating. Each of the three hot legs to the motor is to be protected by a heater-actuated safety overload switch. A 208/24 volt transformer is to be used to provide low voltage power. Draw the wiring schematic for such a system.

DIAGRAM 8

A chiller system has a water pump and a compressor motor rated 208/3/60. A manual pushbutton motor starter is used for the water pump. A flow switch interlocks the compressor operation with the operation of the water pump. The chiller compressor operation is controlled by a line voltage thermostat through a pump-down cycle.

Using the following symbols, draw a schematic for this system.

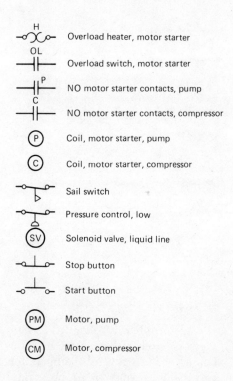

H	Overload heater, motor starter
OL	Overload switch, motor starter
P	NO motor starter contacts, pump
C	NO motor starter contacts, compressor
P	Coil, motor starter, pump
C	Coil, motor starter, compressor
	Sail switch
	Pressure control, low
SV	Solenoid valve, liquid line
	Stop button
	Start button
PM	Motor, pump
CM	Motor, compressor

DIAGRAM 9

Complete the pictorial diagram shown in Figure B-2. The 115-volt power supply is connected to the primary of the transformer and also to the fan motor through a temperature actuated fan switch. The secondary side of the transformer provides power to the gas valve through a limit switch, pilot safety switch, and remote mounted heating thermostat.

Draw a ladder schematic of this system.

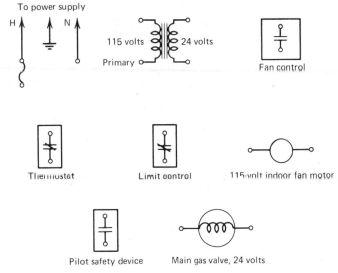

FIGURE B-2

DIAGRAM 10

A line voltage relay coil is to be energized by a heating thermostat. This relay will in turn energize two fan motors in sequence. The second fan motor will be delayed in starting until the first comes up to speed and causes a sail switch to close.

There are three errors in Figure B-3. How would you correct them to provide the desired operation?

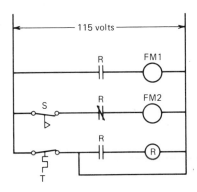

Legend:

   R = Relay
FM = Fan motor
   S = Sail switch
   T = Thermostat

FIGURE B-3

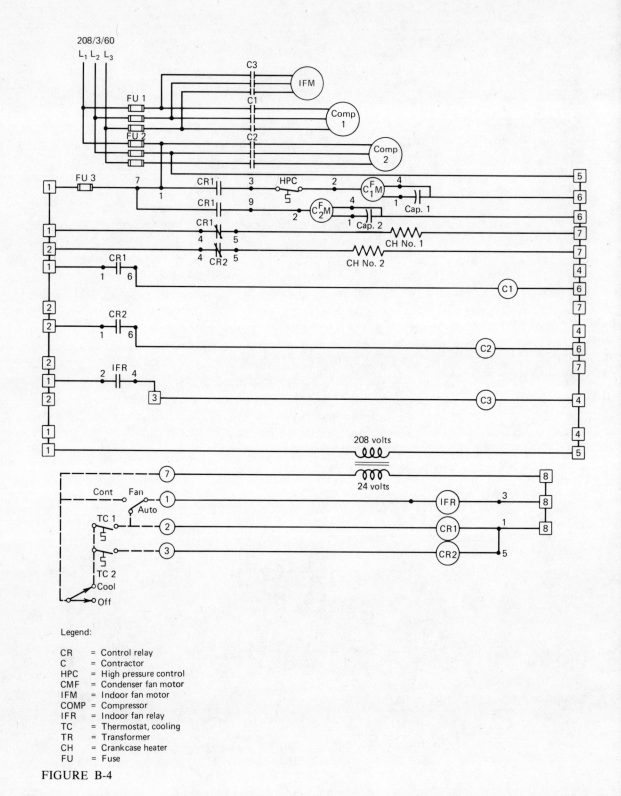

FIGURE B-4

354

DIAGRAM 11

Study the wiring schematic in Figure B-4 and then answer the following questions:

1. What is the rated voltage of each of the following parts?
   IFM
   Comp 1
   CFM
   CH
   IFR coil
   CR1 coil
2. Describe the sequence of operation on a call for cooling when TC1 closes.
3. Under what condition does CH No. 2 become deenergized?
4. If FU3 (fuse) were to blow out while the system was running, what would happen?
5. Describe the sequence of operation when the fan switch is moved to the "CONT" position.

DIAGRAM 12

Study the wiring diagram of the fan center in Figure B-5 and then answer the following questions:

1. Describe the sequence of operation on a call for cooling.
2. Describe the sequence of operation on a call for heating.
3. Would a burned-out coil have any effect on the heating system operation?
4. Would the heating system work if the transformer was burned out?
5. Redraw this diagram as a ladder schematic.

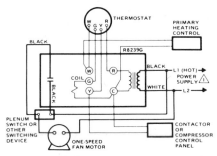

FIGURE B-5

Index